The Conestoga Compact

A Composite Biography of Adam Thorne

By

David Jacobic

Table of Contents

Forward

The name Adam Thorne is now, of course, generally known to nearly everyone on Earth having any sort of average knowledge of the affairs of humankind. He and his compatriots, having accomplished one of this species most significant achievements to date, will join the ranks of the world's most revered, shall we say, explorers. From the mold of the first humans to ever explore what lied past the next horizon, the first to build boats, the first to sail oceans, or the first crossers of the Bering strait, Thorne and his companions cannot help but be known to all human inhabitants of this planet for as long as we care to remember ourselves. Mountains of material have already been written about them, but I must admit that none of this has been authored by anyone as close to them as I have been. I am not a true author in the sense that I have not studied literature or the art of authorship in any way during my lifetime. I am an engineer by trade, and most recently completed a long career as an aerospace designer. I took over from Adam as Chief Executive Officer of Uplift Aeronautics, which is still a quite viable corporation known well worldwide. I was hired by Adam's father, Frances Thorne, in the company's earliest days. I happen to be 31 years older than Adam, and I was fortunate enough to meet him when he began accompanying his father to work around our company grounds. He was a very bright kid, as his father had been, and Adam spent a very large part of his younger years alongside his Dad. The two wound up being more like close friends than a parent and child. They were both quite brilliant, and though Adam didn't take quite as well to the formalized education available to him here in southern New Mexico, it was generally because he was thinking past his teachers. It was my good fortune to spend a great deal of time with the young man while he was at work with his father. When his father passed away at the company factory (from a chemical leak, famously), Adam was already prepared and able to assume full ownership and management duties. I was assigned as Deputy Manager of Operations and was actively involved in all progressive moves the company made thereafter. This generated all manner of innovations, as will be outlined in this book.

I am calling this work a "Composite Biography" because it originates from multiple sources. Much of it comes from my personal memories, as I lived through many of these events myself, and have fortunately been unable to forget any of them. As I am trying to present this material in the form of a novel, I often need to resort to referring to myself in the third person, which has been rather awkward. Please forgive me if it reads strangely to you as well. As for sources, quite conveniently Adam had an unusual proclivity to record all company meetings and conferences he attended at his company and save those records for review. He did this so that he could recall what decisions were made in order to make sure those decisions were the best. He was the type of manager who didn't believe himself to be infallible. Quite the opposite, if someone had put forward a better idea than he did, he would get their permission to use that idea. I have used much of the content of these meetings to describe the progression of events that led to the accomplishments of which we all are so aware. For my final source, I am privileged to say that Adam and his crew, yes, I will call them that, have had the foresight to arrange a means to continue to communicate with us from their progressively more remote location. It was one of his primary goals to ensure that this would be possible. A large portion of this company's resources, both financial and technological, were devoted to making this a going concern. I have usually been the one to receive and review these communications, and I have tried to share them with the public, via the media, when appropriate. Of course, this duty will now be passed on to my successor.

It is fascinating to me that Adam continues to use his established leadership style even now when he is faced with unprecedented challenges as an "Explorer." For him, setting up a team to accompany him, unquestionably for the rest of his life, was the most vital task he would ever undertake. All through human history he felt, groups of people would band together with the self-assigned task of leaving familiar surroundings and going into places they did not really belong. From Columbus challenging a vast and unknown ocean to the so-called "pilgrims" settling on an unknown continent to Elon Musk depositing a colony on an uninhabitable planet, humans have an unquenchable desire to disperse. In retrospect, all these past endeavors were somewhat ill-advised, and should have been done with much more pre-contemplation. But these things were done. Not theoretically, but in reality. So has it happened once more. Adam at least had the foresight to know that only very unusual people were suitable for such an undertaking, and he spent as much effort finding the best individuals to help him reach his goals. We see this kind of screening process again and again in motion pictures, but to do it in the real world is far more challenging. From what we know so far, he did this job quite well.

Since this story will come from such a wide variety of sources, not to mention viewpoints, I will try to tell the tale as a novelist would. This being the case, I have generally chosen to tell the story in the third person. This includes narration in which I myself am the subject. Not being a novelist, as I have said. This book may tend to sound more like a technical manual than most readers might prefer. It is probably inescapable. Previous books written by astronauts, engineers, and the like cannot help but fall into this literary trap. But this is, after all, a biography about an engineer. This is science, not science fiction. But it is even more about humanity and what it means to be a member of this tremendously peculiar species. Also, the story is not over, and there is no knowing how long it may go on. It may easily be true that we will have no idea of the outcome in our own lifetimes, or even in the lifetimes of our children. But whatever facts come to light, and whenever that is, someone like me will make them known.

Dr. Nathaniel Floatingfeather

Chapter One
Foundations

From the personal recollections of Nathaniel Floatingfeather….

In the early morning hours on Monday, August 16th, 2079, the faculty parking lot at the Thorne School of Aeronautical Engineering saw a distinctive car arrive that had not had the occasion to park there before. The space was simply labeled "# 38, Faculty Only" and was completely identical to all the spaces surrounding it save for the number. There was a security gate at the lot entrance, and the driver obviously knew the entrance code when he drove up and punched it in. Dozens of faculty members would follow as this first day of classes commenced for the start of the 2076 fall semester. If anyone were really around to pay attention, they would have been somewhat curious to see that this newbie automobile was in fact a model 2077 Maserati 2-door sports car painted bright yellow, with a white racing stripe across the middle. It was egregiously new and shiny, and rumbled pleasantly as it idled at the gate. The passengers also seemed somewhat out of the ordinary. The driver was a young but very solid-looking young man wearing a bright, blue-colored turban, and a short-sleeved light blue shirt, which veteran New Mexicans could easily identify as being the style of Sikh ethnic groups common in several regions of the State. The passenger was a young man of perhaps 18 years old, about 5 feet ten inches, and slightly slender build. His hair was light brown. He wore blue jeans, gray sneakers, and a Buzz Lightyear T-shirt with "To Infinity and Beyond" written underneath. The two parked the sports car, grabbed a pair of backpacks from behind the pull-down seats, locked up, and shuffled into the building. Fortunately, from a privacy standpoint, the parking lot had high walls around it, and not a soul was around to notice.

The Thorne School of Aeronautical Engineering was the newest college to be added to the other various sub-colleges at the University of New Mexico, located in Albuquerque, the state's largest city. Established in the early 1900s in a single stuccoed building, it enjoyed a steady growth as the fledgling state's largest institution of higher learning. It held the State's only medical school, a solid School of Law, a highly regarded school of Anthropology, and much else.

Also of high regard, the various Colleges of Engineering, encompassing Civil, Electrical, and even Nuclear studies, produced highly qualified graduates in large numbers each and every year. The State, already known for the locus of development of nuclear energy, and of course weaponry, was an ever-expanding location for computer technology and, notably, aerospace science. Much of this remained centered in the more northerly Los Alamos, with its massive National Laboratories. In the early part of the 21st century, though, a shift occurred to a small degree when New Mexico invested in a small, remote piece of real estate near the oddly named small town of "Truth or Consequences" was chosen by a certain Mr. Branson as a locale to build a "Spaceport" to accommodate his Virgin Galactic suborbital spacecraft, called simply "Spaceship 1." These were the early days of civilian space exploration, tourism really. These days seemed entirely reminiscent of the very early days of aircraft development, with wonky engines and wooden airframes, and pilots with goggles and leather helmets. These heady days gave rise to all the future designers and companies that soon developed the warbirds eventually even airliners that still operate today. In just the same way, the mega-billionaires at the turn of the millennium latched on to the idea of space travel as the context for the new barnstormers of the entire planet.

Of course, as technology zoomed forward, armed with new materials that could be fabricated on 3D printers and astonishing new supercomputers to aid them in designing these new types of vehicles, it was inevitable that visionary minds would attach themselves to space travel. New Mexico was thought of as appealing because it was remote, and eager for money and talent from outside the state. Frankly, it was also thought of as easily exploitable, and large companies were more than eager to seek out the tax breaks they could receive, and the low wages they could pay employees when compared to other states. "Spaceport America," as it was formerly called, took nearly a decade of development and lots of public funding to get off the ground, in the literal sense. But by the 2020s, a steady stream of very wealthy clients began dishing out the funds for a remarkably brief suborbital journey to see the curvature of the Earth. This was useful only as a flashy accomplishment that they could use to feather their egotistical caps. Profits increased exponentially, and so did the development of the spaceport. As Branson hoped, his very long runway soon had some very interesting startup companies paying him for hanger space. Truth or Consequences began to actually be consequential. Though the larger city of Las Cruces, less than an hour to the south, would be the primary residence of those who worked at the Spaceport per se, "T or C," as it is usually called, played host to increasing throngs of tourists and other spectators, and the ultimate joyriders, demanded. Restaurants, hotels, and even casinos started rising from the semi-desert. The Rio Grande River, which was barely a creek as it passed through this region, had been dammed in the mid-1900s to give rise to the state's largest lake. This had the equally incongruous name of "Elephant Butte." The lake was not far from town, and the existing marina there was expanded to keep pace with the town. Fishing the Elephant was a bit of a draw, as was boating. But the newly erected resort proved to be the main draw for the ever-expanding set of technicians and scientists working every day at the Spaceport.

Just as the Elon Musks and Jeff Bezoses of the world hired legions of Aerospace Engineers to build the new Empires of the Air, so too did Branson. It is doubtful whether the now very well-aged billionaire took any notice when, in 2051, a teenage boy applied for a job at the Spaceport Terminal in whatever capacity he could get as a summer job. He lived in Las Cruces, and his mother needed him to find some work following her divorce from his father, and money had gotten considerably tight for her. Spaceport was a common place to seek employment for teenagers in southern New Mexico. This was considered comparable to working in an amusement park or even a movie theater in Cruces at the time. The pay was a bit higher than those types of venues, though, and the young man was quite pleased to be hired on as a janitor and sometimes groundskeeper. This youngster's name was Frances Thorne. He was 16 years old and a junior at Las Cruces High School. He was bilingual, as his father spoke English, and his mother, Josephina, both Spanish and English. This distinctive dialect was known as Spanglish, and is still in use today. The ability to wind your way through both languages simultaneously was, and remains, a valued talent in southern New Mexico.

To his own delight, young Frances came to be appreciated rather quickly at the expanding spaceport. He was uncommonly energetic and seemed to find ways to make himself involved in all manner of necessary odd jobs. He would repair leaking faucets in restrooms, for example, without asking permission. He made it his business to know where parts and resources were around the facility. When he was caught having done some unauthorized repair or other, he would ask his irate supervisors if he should go back and break them again. In the sort of workplace he had chosen, the talented and can-do types of people he worked for were quick to notice young people who were cut from the same type of cloth as their spacesuits. By July, has was assigned to work on the flight line, assisting with equipment moving, fueling operations, and even spacecraft maintenance. He held up well under scrutiny and actually enjoyed the attention. When the school year came back around, he was asked to continue working on the weekends. After a year of employment, he already seemed to be something of a fixture there. He was often asked to lead tours of the facilities, especially for Spanish-speaking visitors. He was appreciated as a knowledgeable and easily understood guide.

As he progressed through high school, his superiors could not help but be interested in his progress. It was obvious that he was capable of comprehending the specialized lingo of aerospace and its challenges as well. He found himself assisting in the hangers and watching the mechanics doing maintenance and repair. In school, he had joined the Air Force Junior ROTC unit there and was thriving as a cadet. His reports for the Aerospace education portion of the program continually astounded his instructor, a retired Colonel who had commanded an aircraft support squadron. Fortunately, Frances knew never to disclose any trade secrets, but he obviously knew what he was talking about in class. The old Colonel was soon disappointed to learn that his star cadet did not intend to enter the armed forces, Air Force or Space Force. His Spaceport mentors had told him he had other opportunities he might explore. By his Senior year, he was all but adopted by the young supervisor of Spaceport Spaceflight Operations, a man named Nathaniel Floatingfeather. Nathaniel was a member of the Apache nation, and people often remarked that his name was rather appropriate for such an aficionado of flight. A recent PhD graduate of MIT in Aerospace Design, he recognized talent whenever it presented itself there at Virgin Galactic. Hoping to secure

Frances' talent for the company's use, he helped the young semi-janitor gain a full scholarship to his alma mater. France's high SAT scores, especially in mathematics and physics, were of great help in this effort. His mother was quite flabbergasted, he later said, but bade him an enthusiastic farewell when he left for Boston. New Mexico State, located in Las Cruces, would later say they missed their chance to gain a promising student.

So off went Frances, now accepting of the name Frank, to study at MIT. Obviously determined to get his bachelor's degree in aerospace engineering, he apparently adjusted quite well. He loved being around the same types of fellow students and professors that he had encountered in T or C. Not found to be particularly outstanding at first, he eventually was discovered to immediately absorb everything the curriculum could throw at him. His grades were nearly perfect. The possible exception was Calculus, where he was just a bit below absolute perfection. In later life, he stressed that mathematics was the purview of supercomputers, on which he entirely relied as soon as they became available to him. His undergraduate time was completed in three years, only moderately unusual at MIT. The following year, 2057, he completed his Master's in Rocket Engine Design, focusing on the still rather obscure Aerospike type. This variation of rocket thruster had been envisioned in the 1980s but had never fully been developed for a number of reasons. It was thought to be inconveniently complex and was felt to be at high risk of just plain melting down from the intense heat which always accompanied the combustion of highly volatile fuels. His thesis was met with some interest, and with continuing financial aid from his New Mexico Benefactors, who endorsed his applications for scholarships, he spent a further two years completing a PhD of his own.

His dissertation was what truly caught the attention of the world of Aerospace. In it, he outlined the blueprints of what has since been considered the next step towards greatly increased access to Earth orbit at affordable prices. He designed an entirely reusable and highly capable spacecraft and put it all in actual blueprint form. By merely expanding theoretical principles discussed for years but thought unworkable, he worked through a myriad of solutions to make his plans work in reality. He incorporated the ever-expanding principles of 3-D printing and materials design to conquer the heat management problem. He made his spacecraft roughly in the shape of a so-called "Lifting Body," or a triangular wedge shape with the nose at the narrow end. This was another often recommended but seldom utilized idea. The craft would use Liquid Hydrogen and Liquid Oxygen as fuel. The heat shield would cover the entire belly of the craft, as with the famous Space Shuttle that NASA had used to build its Space Station. However, this shield was made of composite materials, as were the famous glued-on tiles the Shuttle used, often to fatal results. As a replacement, Adam's shield would be 3D printed as a single piece, save for the landing gear covers, and sturdily bolted to the bottom of the spacecraft. The whole thing could be, and usually would be, removed after each flight and be replaced by another. The old one could, in all probability, be refurbished or recycled. In addition, retractable jet engines could be deployed from the top part of the ship after its initial reentry phase to allow for a better-controlled landing cycle rather than a mere glide pattern as was used with the old shuttles. This was another old but unused idea.

Frank realized that to contain all of these features, the ship would need to be rather larger than those that had come before. This being the case, why not provide a larger cargo capacity. This was of course the real reason for such a design in the first place. Humanity had already begun the practice of launching all manner of satellites and short-term research platforms into orbit, not to mention tourist flights and full-on space stations by various nations. Elon Musk was launching his large rockets all the way to Mars as a matter of routine. Spaceflight had progressed from the barnstorming age to at least the days of the Ford Trimotor. Frances Thorne's dream was to try and propel things to the age of the larger cargo planes. The ship he envisioned could haul twice as much cargo as the old space shuttles but be capable of launching from a conventional airstrip like a conventional modern jet airplane. Using Aerospike engines in a row at the rear of the craft could be gimballed to reach any orbit. They could perform as needed in orbit and discharge cargos there. Then turn around in flight and come straight back home again, using only a runway to land. Also, it was more reusable than any prior spacecraft. Production costs for such a craft would also be greatly reduced. Such are the ideas that successful dissertations are made of.

Of course, the publication of this paper caused as much commotion as one might expect. Fortunately, and heavy on the fortune part, our young Frank, only 24 years old, had had the foresight to apply for and receive a patent on his design prior to submitting his dissertation. Although a myriad of designers had speculated and theorized about many of the features of his new space plane (Encompassed, for example, in a NASA idea called the Venture Star as far back as the 1980s), Frank Thorne alone had transformed these ideas into a truly workable design. Even before receiving his Doctorate diploma, he announced that he would soon begin selling stocks for a new company dedicated to manufacturing his new orbital spaceships. The company would be given the now legendary name of "Uplift Aeronautics." He decided to call the fledgling ships "Shrikes," a small but aggressively predatory little species of bird. He applied and got a permit to set up flight operations at his beloved Spaceport USA. His company would be entirely independent of Virgin Galactic, but the State of New

Mexico actually owned the facility, and they were more than happy to welcome him home. As the money began to flow in from Uplift's extremely lucrative stock sales, he set up a Shrike production plant in Grants, NM, west of Albuquerque on Interstate 40. New Mexico state government was ecstatic. The often-forgotten state had been a cradle for numerous mega businesses over the years, including Microsoft and even Amazon. But those promising toddlers had sprinted away from the crib as soon as they could stand upright. Not so this time.

As stock sales began to skyrocket, Frank soon found himself able to start constructing the company he had envisioned. Land was acquired, permits were issued, and the 3D printing equipment and materials he had chosen were acquired within a year. His staff was carefully selected after intensive screening. His young former mentor, Nathaniel Floatingfeather, was an easy choice for director of development. A member of the Mescalero Apache tribe, Frank knew this was the man he needed to transform his blueprints into tangible, buildable substance. The acreage needed in Grants was purchased at a low, state-subsidized price. Utilities and support systems installed, and a large new plant was erected in short order. New Mexico citizens were given first priority in hiring, but professionals from all over the country and the world began to converge on previously poor and obscure Grants, NM and Truth or Consequences as well. Incredibly, the manufacturing plant was ready to begin production by the middle of 2061. Simultaneously, a new management, research, and development facility, ten stories high and fifteen thousand square feet in volume was erected to the east of the original Spaceport USA grounds. This was completed just a few months prior to the production plant. A few dozen miles to the east of this was, and still remains, White Sands missile range. This US government facility for rocket and defense research dates back to the early days of defense missile development. Incidentally, this also encompassed the area where

Trinity site was located, the test site for the first atomic bomb. The area was already restricted airspace due to the experimental phases of Richard Branson's reusable rocket launches for some time. The airstrip was now greatly expanded for the Shrikes to use, and a storage and repair hanger of massive volume was likewise erected near the flight line.

Frank Thorne realized, without much foresight required, that his spacecraft, if expected performance parameters were achieved, would have too many uses to even envision. For example, satellite placement and retrieval would be one of its mainstays. He would also propose that the Shrikes could begin collecting "Space Junk," floating debris from missions dating back to the 1970s that had become floating hazards too numerous to count. NORAD, the government Agency charged with tracking all the debris, was amazed that some huge orbital apocalypse had not yet ever occurred. Frances hoped that Shrikes would become the new street sweepers of the sky. Additionally, construction of new orbital facilities, from observation stations to mega-high-priced hotels, would need cargo space to hoist them aloft and then assemble them. These new vehicles should fit this and numerous other bills. Much less expensive than other lifting options, Uplift Aeronautics and its fleet of orbital craft ought to be in high demand. Resulting profits should then enable exponential expansion, and thus, the novel new spacecraft would pay for themselves. By November of 2061, the first Shrike was completed. By April of 2062, that first one was FAA certified, and the first orbital flight and return were completed by June 2nd of that year.

Frank was very concerned that his new operations were, to use the nomenclature of the time, "Carbon Neutral." As his Shrikes ran on Hydrogen and Oxygen, the most sensible option was the splitting of the $H2O$ molecule into its two constituent molecules to make the necessary H2 and O2. New Mexico had gone to great lengths to become a national hub of Hydrogen production but had at first opted to do this by extracting the Hydrogen from underground natural gas. Unfortunately, this resulted in CO2 production as a by-product, and the entire strategy was to cause great consternation among environmentalists. The erroneous solution was to collect the CO2, and to pump it back underground to increase the pressure in the cavities the gas had come from in the first place. This strategy wound up making the gas extraction that much faster. Of course, even more CO2 leakage just could not be avoided, and the world stayed in the same pickle as it was using fossil fuels for energy in the first place. Uplift, Frank, that is, realized that using Wind and Solar to generate the electricity to split water molecules was by far the best option for generating the gases he needed (in liquid form), to power his ships. Wind and solar were certainly abundant energy sources available in New Mexico. These were available, in fact, for more days every year than just about anywhere else. Unfortunately, the truly crucial element, water, really was not. This was perhaps the most perplexing problem Uplift faced amongst all its innumerable challenges. If a water-rich environment was the primary factor, Michigan would have become a better place to be. If not produced in New Mexico, these two supercooled liquid and explosive elements would need to be shipped over great distances, either by truck or by train. Obviously, this was hardly an option for a man who was entirely dedicated to operations in a desert. Remarkably, fate intervened in that the Aquifer under T or C had apparently replenished itself of late. Perhaps the worldwide shift away from fossil fuels and towards renewable energy had helped the climate refresh itself somewhat. But whatever the reason, longstanding drought conditions in New Mexico had alleviated to the extent that it was projected that Uplift could pilfer underground water for its own use, at least for the foreseeable future.

The flow rate had also risen in the Rio Grande River, which happens to gurgle right through the middle of Truth or Consequences. A new, low-loss fuel production plant was built very near the operations facility. Things were falling rapidly into place.

For such a massive undertaking, this now well-known business concern called Uplift Aeronautics performed a true miracle by making itself exceptionally close to being operational in just a bit over two years. Technological innovations begat technologic innovations at astonishing speed. Frank Thorne felt deep down in his imaginative bones that the state of the human race and human ingenuity was ready for its next lurch forward. This new product, orbital accessibility, was now widely available to more people than ever before. The demand only increased exponentially, and this is still true today. Within ten years, over 150 Shrikes were to be produced, and all were flown out of Spaceport USA. Of course, the name has since been changed to Uplift America Station, a household name around the globe. Shrike based spaceports have now been built in seven countries, and several more are begging for them. NORAD headquarters has required a few more supercomputers, not for tracking space junk, but for tracking Shrikes. On the day of that first successful flight, Frances Thorne threw a massive celebration for his production team at the Elephant Butte Resort and Casino. Not even those brilliant designers could imagine the extent of their contribution to human history that started that day.

Chapter 2
Learning to Fly, Learning to Walk

From the personal reminiscences of Nathaniel Floatingfeather and other Uplift personnel…

By the fall of 2062, enough of the infrastructure at Uplift America was in place that formal operations could commence. Offers started streaming in at once, and orbital flights began occurring about four times a day. The US government demonstrated a willingness to offer contracts for cleaning up space debris in the orbital plane of the International Space Station. Plans were soon made to begin manufacturing new modules for the station itself, which was by this time bordering on becoming derelict. The Shrikes had, obviously, come installed with manipulator arms and other equipment conducive to this sort of work. The forward crew compartments on the Shrikes were more spacious than those of the original space shuttles, and all included an airlock to the storage bay in the forward section, not in the bay itself. The crew compartment had launch seats and living spaces to accommodate a crew of six for up to two weeks. Flight control and electrical equipment were in the form of flat screen and heads-up display types, and onboard Computing and Navigation were orders of magnitude smaller and lighter than the mid-20th century could ever have provided. All of this boded exceptionally well for the fledgling new company. The most important element of all of this was, first and foremost, finding the personnel required to fly the birds. Pilots and flight specialists were needed in numbers unheard of before Uplift America was ever conceived.

As with all the various Space Programs that had appeared around the world before this time, Uplift found itself reliant on former military personnel. The US Space Force was a primary pilot source, but all other services made their contributions as well. Even the US Coast Guard had a few aeronauts to offer. As flight controls and equipment were developed and manufactured, so too were flight simulators. Also, as early as the 2010s, artificial intelligence computing had advanced rapidly, allowing by this time for the Shrikes themselves to handle the computation of courses, orbit changes, and docking maneuvers. Out of old habits, takeoffs and landings were done by the pilots but could be done automatically if the need arose. Hundreds of Shrikes required thousands of flight crew personnel, and part of the large Uplift production facility in Gallup was dedicated to producing the human part of the equation. This was where the majority of flight and ground crew training was conducted. Control tower and ground monitoring personnel were trained there too. So many flights would require NASA like mission control capabilities multiplied dozens of times over. A huge supercomputer analysis capability was needed just to assist the assistants. This was located in an underground facility near the launching site. Frank Thorne and his top-level managers oversaw the construction of all of these vital departments. Every penny of revenue generated from stock sales and pre-commissions was required to put this huge network in place. The collected governments with national space programs on Planet Earth, if combined together, could have equaled such a task.

Although his own personal wealth was increasing quite rapidly, Frank himself continued to live frugally. He had a modest but well-appointed home built near Elephant Butte Lake and drove a brand-new yellow Corvette to work each and every day. He was quite well versed by now in the history of American Space Flight and could spout off the personal biographies of American astronauts throughout the entire Apollo program. This was one of his true obsessions. Corvettes were a chosen automobile for many of these idolized super-pilots. The Apollo 12 crew of Conrad-Bean-Gordon apparently had even had a matching set of three with their names emblazoned on the sides. Frank would zip to and from work, even on his days off, and would pay the numerous traffic citations without a second thought. Oddly, local judges never once revoked his driver's license. When the pilot training center in Grants was first opened, he made the obvious decision to include himself as one of the first students. His justification for this was similar to that used by Deke Slayton when he approved Alan Shepard to command Apollo 14 even though Shepard had been sidelined for years with inner ear problems. "He knew Shepard could do it," Frank said, "And I knew I could do it too." Besides, who could ever know more about the Shrike than the man who designed it. It was as if Werner von Braun had become an astronaut.

Frank began crewing duties about two flights a month six months into the initial operations at Uplift. He started commanding missions two months after that. Usually, these flights were particularly important ones, the ones that set new precedents in pioneering new technologies or equipment. He often made the nightly news with his exploits. New contracts came quickly, and every Shrike on the tarmac became an essential bird. Truth or Consequences was glad that the launch facility was far enough away that all the noise was no worse for them than that of a modern airport. In fact, the Shrikes didn't power up past the speed of sound until they were well over the Caribbean Sea. About a year into operations, Frank had taken in enough in commissions himself to actually buy out the company. Though not the sole owner, he was certainly the controlling partner. He likewise became a billionaire many times over and, by then, was one of the top ten wealthiest men in America. He had a new swanky mansion built for himself east of his spaceport, just at the eastern edge of White Sands Missile range. His Shrikes began flying over his property multiple times every day. He didn't mind, though, as he was hardly ever home. But he did make sure to have the place made very sturdy and soundproof, almost a gilded age bunker. It has never suffered any damage at all, right up to the present day. He also refrained from naming the estate with some pretentious moniker. He just called it "my house." There was a swimming pool there out in the desert and a garage befitting a lover of the high-tech vehicles he kept buying but seldom got to use.

His life to that point, having been now substantially established there in the New Mexico desert, was filled with constant labor. But a young man is a young man, and Frank found himself a severely wealthy young man with no sort of social life whatsoever. The gossip press began to seek him out, inevitably hoping to find some scandal or other to expose or some sort of love life into which they could interfere. Frank did happen to be moderately handsome as well, so it followed that he would wish to commence a love life eventually. It always seemed to be the case that the super-rich of America, and really all wealthy nations, were hounded by women constantly throughout their lives. Our young Frank could not avoid it. Neither did the jet setters of the world, especially the young wealthy ladies, intend to let him avoid it. As a basic introvert, he never threw parties or social gatherings at his mansion unless the guests were fellow employees at Uplift. But what billionaire could escape trips to conferences and gatherings out of town? The upper crust types in New York, Los Angeles, and everywhere subcontractors dwelled clamored for his attention on a fairly regular basis. If word got around that he was around, politicians and celebrities had a way of squeezing through the door to see what just what he was about. Via these innumerable channels, several dynamic young ladies began to vie for his attention. And among these, one actually managed to eventually catch his undivided attention. This, of course, was, as everyone knows, the universally adored actress Verity Lovato, a veteran by this time of more than a few motion pictures. Accustomed to obtaining whatever goals she set in life, catching a billionaire husband was definitely goal number one.

Now, just like other famous celebrity pairings in modern history, Marilyn Monroe and Joe DiMaggio for example, the match was not at all devoid of love. The two were both extremely fascinating to each other, and both immensely interesting in their own rights. Verity thought Frank's life must be interesting indeed, and Frank thought Verity must be mysterious and alluring if he looked close enough. Neither of them were particularly bad people, not at all. Whatever problems arose later on in their lives were due to the fact that they were, especially busy people and didn't want to give up what they were doing to make time for themselves. They each needed to be in love simply because they wanted to experience it at least once in their lives, just to see what it was like. That goal, like all the others they set for themselves individually, was fulfilled very sincerely until the time it was met, and then it was discarded without a thought. They paired and then were as inseparable as the world allowed. They were married in Paris within three months of the first encounter. Verity then went to live with Frank in his mansion in the desert. Frank was only away from full-time work for a total of about a month for the honeymoon period but still took calls in this study. Verity was also at home for three months, but only because she hadn't accepted any acting jobs. The tabloids buzzed away, just as the Shrikes kept buzzing overhead.

That both of them wanted children was never in doubt. Frank was eager in his heart of hearts to pass on the legacy of care and encouragement that his mother had passed to him. Verity, as her last name suggested, was of Latin American descent, as was Frank's mother. Only Frank's father was of European descent. Not to say that Latin Americans were more apt to procreate than any other group, but it seemed that Frank especially wanted this legacy to continue, and if so, it should be continued in the land of his birth, New Mexico. He wanted his children to remain bilingual in Spanish as well as English, as both he and his bride were. And true enough, within a year, in late 2061, there came into this world a healthy young baby boy. For whatever reasons the parents had, the very young man was christened Adam. Perhaps it seemed ordained that as a firstborn child, the first name of the first human being seemed appropriate. With Adam's birth, Frank was as ecstatic, perhaps, as God himself must have been on the sixth day. Verity was certainly pleased as well, and she stayed at home at the Spaceport for two entire years without accepting any work. She and Frank were often inseparable, even at the Uplift facility, and people began to wonder if she was to be a Yoko Ono of spaceflight. Verity, at least generally, did not attempt to inject herself into any company decision-making, as many feared she would. She also continually declined any offers to accompany her husband on any orbital flights. She forbade her young son to do so as well.

After two years of this bliss, Verity began to entertain the idea of returning to the Motion Picture industry. She had no need for the money, obviously, but she was, in fact, a very talented actor, and various studios were vehemently demanding her return to the trade. Even if she had been offered a royal title in some monarchal nation, she would have turned it down in favor of her beloved profession of acting. She accepted a female lead role in a remake of "Ben Hur" and needed to leave New Mexico behind for the better part of seven months. This was ironic since "Ben Hur" had been written by a fellow named Lew Wallace. Wallace was a retired Civil War officer who had penned this masterpiece in Sante Fe during his time there as territorial governor. She did fly home intermittently to visit her two young men, but otherwise, she became a mostly vague figure to her infant son. Frank made up for this by starting a daycare center just down the hall from his office at Uplift. He always managed to spend time with his idolized son, who was banned only from office meetings and visits to dangerous work areas. At home, he apparently loved the sound of launching spacecraft as he heard them through the mostly soundproof walls of his nursery. If it were three in the morning, he wouldn't be awake crying. His father would hear him through the baby monitor giggling and yelling "Zoom!" very distinctly. From infancy through his preschool years, young Adam was raised entirely by his father with the help of the on-site nursery and then the company kindergarten.

It seems unlikely that any human being started life as close to the aerospace universe as little Adam did. Even the Robinson children from "Lost in Space" were a bit further on in age when they left the Earth, and even then, they really never even left the planet when filming. There was always something going on around the youngster to hold his interest. He never seemed overwhelmed by anything Uplift had in store for him. Being hungry or needing a diaper change were his only sources of annoyance. As time went by, his mother continued to be a busy Hollywood film star and would barely finish a film before another one started. She occasionally expressed an interest, Frank told his few confidants, in bringing Adam along to film sites under the care of hired nannies. Frank would have none of that though. Never. Where Verity could be away for months, Frank could hardly stand to be separated from Adam for more than a few hours. Adam appeared to have two parents who turned out to be polar opposites. On those rare occasions when Mommy was home, Frank was forced to confront her with his concerns over her mothering skills. She somehow managed to keep his discontent under control until Adam turned six. Frank informed Verity about that time that, firstly, he wanted more children, and secondly that he would expect her to stay home to help raise them. Quite predictably, she thought for only a few seconds before responding that she actually hadn't understood what being a mother really entailed when she agreed to it in the first place. In other words, it was fun while it lasted, but other things seemed even more fun now. She suggested a divorce for the meager-seeming price of one billion dollars. Frank responded that as a mother, she wasn't worth a plug nickel, but he would give her 500 million dollars anyway. She agreed. The arrangements were made, checks were written, and Presto!, the two never saw each other again. At Uplift Aerospace, the huge staff that had considered itself Frank's family all along were all entirely relieved. Frank astonishingly seemed to put it all out of his mind almost by the next working day. Six-year-old Adam would recall to friends when he had reached High School that anything he ever really knew about his mother came through her movies. As for Frank, he seemed to recognize the folly of having a love for anything except Adam and spaceflight and never ventured into the truly bizarre and imprecise mechanics of love again.

Chapter 3
The Rise of an Original Man

*From recollections of Adam's childhood friends
and school faculty ...*

When Verity departed once and for all, Frank was free to design a path in life for his six-year-old, one and only son. All parents attempt to undertake this task to some extent or other unless circumstances just don't allow it. Frank was quite obviously more than able to provide whatever Adam might need in this direction. He was always laser-focused on giving his son as much opportunity to live a fulfilling life as only a multi-billionaire could. He was also fully aware that as the son of wealth, Adam would be at risk of becoming a target for harm in some fashion, whether through kidnapping, blackmail, extortion, or just the baser instincts of the world at large. At the time, as well as in retrospect, Frank's friends came to realize that he was probably just a bit *too* obsessed in this direction. His concern for Adam's safety was very nearly all-consuming. How much Adam was disadvantaged by this is an ongoing topic of debate. For example, he made sure that Uplift Aeronautics had a significant security presence at its various sites. This was not just for Adam's sake but was meant to address the obvious risks of theft of technology, merchandise, spacecraft, or even terrorism that might come from any points of the compass. Frank had decided early to employ a security firm known as "International Guardians, Inc." which was based in central New Mexico and was owned and operated mainly by members of the Sikh community there. Personnel were highly trained and well-equipped for this type of work. Their reputation and skill were said to equal that of such elite US military units as the Navy Seals or Army Rangers. Known for their wearing of traditional Blue Turbans at all times, they were seen everywhere around the facility, and heavily armed in the areas where it counted. Not all were Sikh, of course, and non-members simply wore blue berets instead. In general, these guards were heavily bearded and built as solid as stone. One could not help but feel safe around them.

After security, Frank's second consideration as Adam started his life was education. As a convenience to his employees, as well as to provide increased safety to their children, he decided to open a school on the spaceport property. This was also somewhat inspired by the system used by the armed forces. Since spaceflight operations were a bit dangerous themselves, given the explosive characteristics of liquid hydrogen as well as the perpetual risk of crashes, the facility itself was several miles away and behind some high hills in the area. There were classes for all grade levels, and the place was simply called "Uplift Academy." There were sports teams and facilities comparable to other public schools in the area, and it was considered to be part of the Truth or Consequences School District. Although reserved for children of Uplift employees, it participated in municipal activities on an equal footing as the other local institutions. Also, T or C taxes did not fund the school in any way. Uplift paid for everything itself. Young Adam started first grade there with other children of company employees. Frank demanded that his son was not to be treated any differently than the other children, be they American citizens or employees from other nations employed by Uplift through work Visas. The school therefore had a very diverse and international flavor. English was the primary language, but not the only one. The school taught numerous language classes and employed numerous translators.

At the age of six, Adam was really quite cute, generally happy, and altogether typical of the other first graders at Uplift Academy. Frank made it a point to tell his teacher, Julia Velasquez, to treat him precisely the same as all her other students. It was made especially clear that he should not be identified as his son. This was for Adam's own good, but the teachers figured out quickly that this was also due to Frank's fears for Adam's safety. Whatever friends Adam made could find out who he was on their own. He generally had his secretary, Frieda Jojola, take Adam to and from school. She drove a Mazda sedan, not one of the expensive hot rods that Frank drove. This was apparently so that no one would figure out that Adam and he were related. As Adam was already very familiar with Frieda, he did not seem to notice that his father was already at work by the time they made the short drive to school. It was strange that Frank thought that these measures would actually make Adam anonymous. No doubt the exact opposite was true. Everyone at Uplift could easily identify Adam on sight, but he wasn't treated differently at school because of who his father was. All that is remembered is that Adam quickly made friends with all his classmates and was thought of as being pretty much the same as everyone else. All students there were the children of Uplift employees and therefore were part of the same family. From technical staff to Spaceflight Ops to eminent scientists, there was no distinction between their children. The youngsters certainly didn't care about things like social standing at that stage in their lives.

As for Ms. Velasquez's first grade class, everyone did remember that there were three kids who seemed especially close. The adults couldn't help but keep a close eye on Adam though, and they observed he had two other six-year-olds who seemed to become his inseparable companions almost from day one. There was a young Sikh boy named Manjeet Singh who was enormously energetic and endlessly gregarious. Adam appeared attracted by his friendliness and playful character, especially at recess. Manjeet was certainly more athletic than Adam but did not appear to mind that his new friend would rather talk and use the swing set than play kickball or shoot mini hoops.

Adam liked the idea that Manjeet generally dressed in more interesting clothes than everyone else. He always thought the same about the numerous Sikh staff he saw around the Space port when he was with his Dad every afternoon. The community was made up of second or third-generation children of immigrants from India and were a familiar presence in New Mexico. Although their native language was still preferred within families and between themselves, all spoke English the rest of the time. They had no perceptible accent other than the usual New Mexico one. At any rate, Adam and Manjeet became truly inseparable. Adam never once in his life ever imagined that the name Manjeet was in any way uncommon or unusual in society.

The third member of the new young trio was a girl by the name of Junaki Sato. Her parents were both design engineers from Japan. Juni was born a bit after their arrival at a hospital in Las Cruces. Her parents both spoke English with heavy Japanese accents, although these would decrease gradually as time went by. They were so enthralled with the work they were doing, and with their own growing importance within the company, that it never occurred to them to return themselves or their daughter to their beloved Nippon. This kind of split nationality situation would become common with employees from other nations over the years, although not universally. Some would wind up returning to their home nations to head up new branches of Uplift Aeronautics or just become homesick and leave. Junaki herself was energetic and forever joyous when around others. She was one of those little girls who had endearing things to say about everyone and everything. She seemed to gravitate to Adam and Manjeet because they seemed as engaged at school as she was. They were always interested in helping her learn whatever it was that was being taught that day in the classroom. She is recalled as being artistic when art was the topic and for paying attention at times when hardly anyone else was. Her friends started calling her Juni, and she didn't seem to mind the contraction. Her parents even adopted the nickname. Manjeet, however, remained Manjeet, and that was because he wanted it that way. He was not a nickname sort of kid and maintained a reputation for severity of demeanor for the rest of his life, so far as we know. Adam simply stayed Adam, but the staff at school had been instructed to refrain from using his last name as much as possible. Even when calling roll in the morning, he was just "Adam."

Since all children age at the same rate, and the student population was smaller than true public schools, the same children all advanced together all through their elementary school careers. New employees would come and go from Uplift, of course, and they brought or took their kids with them, as the case may be. The result was that as the kids advanced from grade to grade it was more a matter of just changing teachers, and perhaps physical classrooms, as everyone aged. Friendships thus forged were therefore not disrupted much during those formative years. The students only grew closer as time passed. This is a fairly universal phenomenon all over the world, but it may have been even more true in a closed-off community like Uplift Station. Adam, Manjeet, and Juni were subjected to this unique isolation to the same extent as all their contemporaries were. Notably, there never seemed to be a dominant member of this particular group of three. Of equal importance, no one of the trio was thought of as a lesser member either. It was inevitable, however, that Frank Thorne's delusions about the anonymity of his son's identity would eventually be shattered. This most probably happened at a student-teacher conference, or perhaps a parent's night or open house. Nobody remembers exactly. Some students' parents undoubtedly saw Frank off by himself and told him that Adam was looking for him. Whatever the case, Frank knew his secret was now exposed at Uplift Academy. In retrospect, it still seems a bit odd that he ever thought it should be a secret. It also seems a bit unusual now how Frank never was known to take Adam out in public. Most Uplift staff generally had homes in the new suburbs of Truth or Consequences or Las Cruces. Although Adam could always invite his friends to the Thorne estate, Adam generally couldn't go visit them. Nor did the two Thornes ever go to a restaurant or a movie in town. They would either rent a movie from a streaming service or order take out. Frank was also a fabulous cook himself. This is not to say they were isolated; they were just very sociable in an isolated setting. If Frank had to leave Uplift for meetings or whatnot, Adam was always left at home.

When Frank finally realized that everyone knew that plain old Adam was actually Adam Thorne, the eight-year-old son of a billionaire, his fears about Adam's safety didn't really fade away. But it ceased to be a major concern as long as his identity wasn't revealed in the public at large. Frances did continue to insist that the school faculty show no favoritism to Adam because of his relationship with the Boss. Surprisingly, two people who never had deduced that Adam's last name was Thorne were Juni and Mangeet. Adam would later recall that these two were more than a bit cheesed off with him for not telling them. This was also the case with many other friends and schoolmates. When word got out, Adam apparently told them then that he was still just Adam, and what they saw was what they got. In other words, forget about it. He wasn't going to change who or what he was, and he didn't expect them to either. Over the years, there were inevitably some classmates who believed they could exploit him into granting favors for them or their parents, but that would never go anywhere. Above all, Frances was always eager to assist the school obtain whatever resources the school might require. Any promotions parents might get at work, or any reprimands for that matter, were due to work performance and not relationships with his son at school. From second grade on, Adam's friends could come over to the Thorne house and interact like any other kids would. Besides, the place was fun and all, and Adam always seemed to know if his classmates were just out to exploit him in some way. You still had to be a genuine friend before he would trust you. He chose his friends carefully, and therefore his father trusted his son's judgment to an extraordinary degree.

By the time Adam's class reached middle school age, certain proclivities were being identified in the young man, and his young friends as well. Teachers informed Frank Thorne that he did indeed exhibit an apparently higher-than-normal IQ. He frustrated them somewhat, however, as he continually jumped past the basic material and questioned how the core knowledge was obtained. This was especially true in mathematics. He was almost certain to ask who had discovered certain math processes, even simple things like long division and fractions, and what thought processes were used to develop such principles. In this way, he resembled his father to a remarkable degree. In fact, he was certainly showing signs of being a true introvert. Of course, these details were generally beyond the knowledge of even the best teachers at mid-school level. In fact, they are generally not even taught until the post-graduate level. Hearing things like "That's just the way it is!" was nearly infuriating to Adam. He would grumble, but then easily grasp the concepts and move on. On one occasion, a teacher pulled Frank aside and suggested that Adam be tested for autism. Frank apparently just laughed, and responded that no, he didn't have autism. What he really suffered from was autonomy. As for Manjeet and Juni, they proved to be more compliant with their assignments as given and remained at the top of their class in every topic. Competition between Adam and his two friends was very fierce indeed. No one of the three was consistently at the very top at a given time, but no other students ever performed as well as they did. There is no evidence at all that they were given special treatment either. Success really was just due to pure initiative on their part. Manjeet and Adam were more generalist in their interests than Juni was. Juni, however, exhibited a near obsession with computer-related topics and the use of related technologies. The three began studying together after school nearly every weekday, either at their various homes or in the small study near Frank Thorne's office in the headquarters building.

Whether or not this access was unfair to the other students is perhaps debatable, but Frank's obsession with supervision of his child's education is a fairly universal phenomenon. To Frank, this seemed the best time of his life. He appeared to enjoy being around children more than anything else he ever did. Manjeet's parents certainly didn't mind having their son at Adam's and, thus, at Frank's elbow. Juni's parents had had frequent contact with Frank for years and considered him a friend even before Juni was born. Manjeet's father was a supervisor in the security department, and his mother a cook in the large facility cafeteria. These were very critical areas and were part of the Uplift community's daily routine. Therefore, they were all quite well known to Frank. What was interesting is that all three members of this mid-school trio were often seen tagging along with Frank as he made his late afternoon rounds. Only Adam got to go into the restricted areas where safety was a concern. Juni and Mangeet had to wait outside. Launch operations were an obvious favorite with every young person at Uplift America Station. Kids were a common sight at the designated observation areas nearly every evening. Frank and his entourage would even visit places like the vehicle maintenance facility, as well as the mission planning and training venues. Employee's children were all fluent in Uplift jargon by early adolescence, but Juni, Adam, and Mangeet were more adept than most. Like all Uplift kids, they all were given daily admonitions to not talk about anything considered sensitive, especially to people they didn't know. Security was severely stressed to Mid-schoolers at Uplift. That was another obsession for Frank Thorne.

Adam, of course, was more of an insider there than any other human being could be. Frank continued to command at least three orbital missions per month. Most of these lasted no more than two days. Frank left Adam in the care of his secretary Frieda on those occasions, although such supervision was barely necessary for the independent young man. Uplift continued its satellite servicing missions at that time, as well as its space junk collection. Such objects were usually returned to Earth for recycling. However, a new innovation now in play involved attaching very small rocket engines to unwanted junk and shooting it off towards the sun. This remains a remarkable but simple bit of innovation. It was decided that a little litter would do no harm to the star, and practice continues to this day. There was also a plan developed to aggressively begin collecting plastic and other "Forever Trash" from all over the world, including oceans, and compacting them into cube-shaped bundles. These too began to be hoisted off the face of the earth and then shot into space. Great efforts were made to recycle the stuff for reuse here on Earth, but the reasoning was, "If we're going off planet anyway, why not?" The small rocket engines used are not expensive, and the Shrikes have since been hired to perform this service from various locations around the world. Plastic production was generally replaced with renewable substitutes in the 2040s, and this clean-up measure for what remains should continue until the massive task is completed. It remains rare for Uplift payloads not to include at least a small bit of sun-bound plastic.

Uplift also began participating in the construction of various orbital facilities. Aside from supplying and upgrading the ISS, the American orbital station (In cooperation with numerous nations), there were orbiting hotel facilities constructed for use by the few who could afford it. These had come and gone since the 2020s. Another project that truly fascinated Frank was a micro-gravity inpatient hospital. Patients there were not charged, and still aren't, as they are considered participants in medical research. There was an interest in the benefits of low gravity for cardio-compromised patients who needed to decrease the workload on their heart muscles. There was research done also on physical therapy for paraplegics and amputees who might strengthen their remaining limbs and enjoy increased mobility while in orbit. The greatest therapeutic benefit of all was for bedbound people who had developed severe bedsores. Despite vigorous turning regimens in their beds at home and countless experiments with specialty mattresses, there were still countless patients whose bone-deep sores would never heal while on Earth. Infections were inevitable with these chronic wounds, and a miserable death would conceivably ensue. It was found these wounds would easily improve in a weightless environment. Granted, transporting these fragile patients into orbit was a tricky business. Launches were extremely stressful for the critically ill, as were landings. Also, developing techniques for total patient care in weightlessness was a huge challenge, but great strides were, and will continue to be made. Such were the activities at Uplift Aerospace while Adam Thorne grew up.

It became inevitable that by the time Adam reached high school at age 15 that the son of Frances Thorne would begin to assert himself in his father's unique sort of world. For example, he developed an interest in the assorted land roaming vehicles his father owned. Specimens of the world's preeminent motor vehicles were in Frank's garage just as Adam was learning to drive. Coincidentally, Frank was always made himself available as a driving instructor. Frank even had a little speedway course constructed on his Uplift property. Adam piloted his cars around the track whenever he could, because why wouldn't he? Though no actual professional race cars were purchased or competitions entered, there were no mundane vehicles at the mansion. Adam developed an intimate understanding of the mechanical characteristics of high-tech cars in a very short span of time. He was again truly his father's son in this respect. Similarly, while growing up next to his father to the extent he did, absorbing the intricacies of space flight was entirely second nature. His father urged him to consider flying with him, and Adam readily agreed. Company policy dictated no passengers younger than 18 would be allowed, but since Frank was the only person who could wave that policy, Adam found himself in a Shrike cockpit jump seat on his 15th birthday. Frank later said he acted like a veteran astronaut from the time the engines lit.

Also, at about this time, Frank decided to proceed with an idea that had been percolating in his mind for several years. As with most international companies, there was a constant need for well-educated professionals to provide the advanced mind power required to perpetuate the goals and activities of whatever the company was doing. Most Uplift engineers, designers, and managers had degrees from major universities around the world. Frank himself was an MIT man, and all the Ivy League institutions and other preeminent engineering schools around the US were represented at Uplift. Name any major university from Albania to Zaire, and there was likely some representation. Not so much was this so from Uplift's home state, though, and this seemed a bit unfair to Frank. This was the State that fostered the atomic age after all, for better or worse. Now it had become a major mover in the space age, surpassing Florida and Texas as a launch point off the planet. Why should outsiders provide all the talent for this effort? Now, Frank never felt that possessing billions of dollars was an end in itself. Doing billions of dollars' worth of good for the world seemed a much better use of one's money. Working with the University of New Mexico in Albuquerque, Frank resolved to add a college of Aerospace Engineering to the other engineering schools already there. This would require a strip of land near the corner of Central and University Boulevard where the other sub-colleges were already located, and some related research lab facilities were just a few miles south near the Air Force base. The University readily agreed with this offer, and especially so since Frank was footing the entire bill for construction. UNM would assist to provide faculty and salaries for those thus employed. Uplift would additionally fund a bevy of full scholarships for talented candidates and would give close consideration to any graduate who desired to work for Uplift Aerospace. Obviously, he had it in mind for his son to attend this institution, and by the time Adam had graduated from high school, the place had begun accepting students. Adam had finished second in his class at Uplift Academy by the time the school opened. Mangeet Singh happened to be first that year. Junaki Sato was third.

All three of these students were offered full scholarships to what would be called the Thorne College of Aerospace Engineering and Science. Frank Thorne spurned the idea of a scholarship for Adam, however. He saw taking such a scholarship as a conflict of interest. He also reasoned that if a student could easily pay his own way, then scholarship money should be saved for those who could not. True to form, he insisted that he be the one to pay for all aspects of Adam's education personally. Also, true to form, while he was at it, he was extremely concerned about Adam's safety. In Albuquerque, it was true he would be relatively close to home, about two hour's drive if Adam managed to obey traffic laws. Adam wanted to live in a dormitory, which sounded like fun. But as it turned out, Frances found a three-bedroom home he could lease in the somewhat swanky Albuquerque Country Club areas not far from UNM. This, of course, would include increased security measures, and a bit more luxury would be added. Now that Frank thought about it, a bodyguard would also be rather desirable. Hmm, now this should be someone who could be by his side as much as possible but also be inconspicuous. But who ever went to class with an armed guard beside him without raising questions? The answer to this question was immediately obvious, as it turned out. Frank realized that all that was necessary was to give Manjeet a bit of training and a few inconspicuous weapons. A keychain baton and a taser disguised as a cigarette lighter were thought to be sufficient. Frank was really thinking this through. It was all coming together! Manjeet was old enough to adopt a traditional blue turban and had a full black beard as well. They had already enrolled Adam and Manjeet in all the same classes. They quickly asked for faculty approval, which was of course granted. Hmm, this did seem a bit conspicuous though. Oh well! Oh yeah, they were also to always be assigned seating close to an exit. Adam never managed to put a kabosh on that idea. Even so, Frank made sure the school entrances were given metal detectors to help screen for weapons, and Uplift security was assigned to all Thorne college facilities. Finally, to satisfy his obsessive father, Adam was asked to come up with a fake name. He decided Adam Thornhill would be sufficient. He told Frank that this would be far too obvious for anyone to second guess it. He thought it was hilarious (in secret), that Frank actually agreed. Finally, Frank was very much gratified that the had actually succeeded in keeping Adam completely out of the press. If the public knew his son Adam was born over 18 years ago, not much had ever been publicized since then. No one was likely to figure out he would be going to UNM.

And so it was that on Monday, August 16th, 2079, a yellow Maserati driven by a stocky young man in a blue turban and a rather plain passenger, arrived at the faculty parking gate and punched in the entry code early in the morning. The Dean had eagerly granted this privilege to the son of his college's benefactor to help ensure his safety. Mangeet and Adam made sure no one was looking as they casually strolled around to the front entrance, winked at the security detail in the lobby, and chatted in adolescent fashion as they made their way to Fundamentals of Aeronautics 101. They sat down in their preassigned seats near the door. They planned to meet Juni for lunch at the student union building at about 12:15. She was over at the Computer Engineering department also starting her first day. Adam's class was not a class she needed. She was afforded no security of her own, but Juni could certainly look out for herself anyway. All three had full freshman course schedules, but the Math, English, and other departments were all aware that they had some new VIP students to deal with. The new freshmen were accepted with no particular fanfare. Frank Thorne had made sure of that.

Chapter Four

School Days

Complied and adapted from diary entries of Adam Thorne transmitted after permanent off-planet residency.

"My life before starting college was really a bit trivial. I was always sort of cordoned off from the rest of the world. Sure, my father and I were not exactly wall flowers when we were at home at Uplift, but he always did his business travel without me. Secrecy was a major deal for everyone working for Uplift Aerospace whenever they were off the grounds anyway. But we Uplift kids all used social media and were certainly far from isolated from society as a whole. Of course, Dad made me use a fake name when I went online. But overall, we kids were probably better informed than most, considering we all had scientists and geniuses as parents. I barely remembered my mother, and most of what I can recall was through what my father told me. We always watched her movies on television whenever they came out, and he made it a point to comment on whatever he liked best about her performances. He never stopped praising how beautiful she was, and how I reminded him of her best characteristics. Then again, he never mentioned anything about how glad he was that she wasn't part of our lives. I figured that part out by myself. It was just a natural thing that she wasn't there with us. He made up for making me live without a mother by playing both roles, I guess. He was always focused on me despite everything else going on. For all of that, I do think he overdid things a little. As a result, I grew up at work, you might say. Any kind of private life was a concept I never knew might exist. My earliest memories were of being with him when he was talking to his people about shipping manifests, flight control, mission planning, maintenance issues, all of it. Work was recreation for him, and fortunately, it was just as interesting and comfortable for me. All of my friends grew up in that environment as well. In retrospect, it seems weird that jumping off into outer space was as routine as taking a day trip out of state was for most other Americans, or an airplane flight to most people around the world. I left the comforts of gravity earlier than most of my friends. But many of their parents were on Shrike crews.

From this isolated microcosm, I ventured out to a different one, the University microcosm. Me and my friends Manjeet Singh and Juni Sato, surrogate siblings really, were provided with a small house in the Albuquerque country club district. It was fixed up very nicely and had great security systems and all the high-tech conveniences any college student could ever dream of. It was about a 20-minute drive from there to school. It also had a well-appointed garage where we could park the infamous yellow Maserati my dad had given me. High end cars were a bit more common in Albuquerque by the time I was there, but it still garnered a bit more attention than maybe it should have. I had my best friend Manjeet, a big Sikh fellow who lived back at the Uplift compound, do the driving. I drove that car myself when back at home, but it was thought to be safer for him to do it in the big city. Damned if I know why. Our other roommate, Juni, was indeed more like my sis, but she shunned our yellow hotrod claptrap and usually took the bus to school. We left for classes a bit early, and usually got home after dark. The neighbors probably didn't like the noisy motor as we came and left, but the garage was locked up pretty tight so that no one could try to steal it or be annoyed by it being there. Electric Maserati's were available, but Dad didn't care for them and neither did I. Fortunately, some gasoline stations were still available nearby, though mainly for ATVs and such. The station closest to our place definitely never seemed to mind when we pulled up. Gas was, and remains, very expensive. Fortunately, we put very few miles on that car, unless we went home for the weekend. Manjeet was funny about never speeding, even though he knew it annoyed me.

The three of us housemates kind of established a rhythm in our daily schedules after just a couple of weeks. We were up about five a.m. and made breakfast and did the dishes. We were all engineer types and preferred our home to be very organized and tidy. (I must concede that I have known a lot of engineers who were very tidy at work but quite sloppy in their personal lives). Manjeet then drove me to school by 7 am, and we took advantage of our status at the college and met in the Aerospace college library, where we had a study room that was on permanent reserve. Juni would generally meet us there as well after taking the bus. There were plenty of times when her computer engineering studies demanded she be elsewhere in the mornings though. Computer people had their own separate buildings. Manjeet, myself, and, before long, others would compare notes every morning and finish any homework or projects we hadn't quite completed yet. In other words, it was a study group. It lasted all throughout four years of undergrad. I unofficially ran the group, and I was the one who would ask others to join. There were other such groups, of course, but our members seemed to consistently get the best grades. It reminded me of Timothy Bottoms and company in "The Paper Chase" film. During this time, no one ever knew that Manjeet and I were operating under false pretenses, or at least no one ever told me if they did. After study group, we all went to our various classes. We all basically had the same Engineering classes together, at least for the first two years. We'd all have lunch at the Student Union building, finish up in the afternoon, then go home. Manjeet, Juni, and I never really socialized with anyone else after school except on a weekend now and then at a bar or dorm party. Juni was the only one who would ever drink or smoke pot, and then only sparingly. Manjeet refrained for religious reasons, I guess. For myself, I considered it a waste of money, and I really hated the way that stuff made me feel. If people were offended by our abstinence, we'd just leave. Our evenings were filled at home with studying, projects, and video games."

"A few of our study mates, of course, knew us from Uplift. They had all taken the "Vow of Silence" to not rat us out before being accepted into the Study Group. That was Dad again. Everyone just referred to us as "Adam" and "Manjeet" and claimed not to be able to recall our last names, that sort of thing. Our study room was Number 9, and that's what we asked for when we reserved it. We started calling our study group "The Nueves" or "The Nines" in Spanish just because we couldn't think of anything better. This was a kind of play on words for "Nuevos" or "New" since we were all freshmen at first. Nuevo Mexicanos were New Mexicans, so sometimes we called ourselves Neuve Mexicanos as well. It's weird how study groups give themselves nicknames, especially when the students are true smartasses. No one we hadn't previously accepted into the group knew we had that room on permanent reserve every morning. It had a large table and excellent Wi-Fi access, as well as a large screen to project whatever was needed at the time. We would generally get coffee and snacks from the cafeteria in the same building, which opened at seven a.m.

While I won't go into everyone's names and stories, as it would be hard to get their permission at this point, there was, of course, one person in the group who would make a major impact on my life. As it happens, she is sitting right next to me as I write this, and she says it's quite alright to talk about her. I met her on the first day of classes, the first class in fact. It was "Principles of Aerodynamics" with Professor Don Prio. Manjeet and I took our tedious assigned seating in the back, another of Dad's ideas. The classroom was relatively large but only about halfway full. As professors usually ask their students to move up to the front, we must have stood out. Our general appearance must have been strange enough. It was even more odd when Prio never had us move to the front. One of our new classmates did object to this, however. She raised her hand to state quite plainly that it made her feel strange to have us back there where no one could see us. Having been warned in advance that this was where we needed to be placed for "safety" and all, he announced that it was nothing to worry about, as we had "medical issues" that he couldn't really reveal out loud. The class didn't seem exactly relieved to hear this for some reason. I gave a short little cough, and class began. I did notice that after raising her hand and objection, I couldn't help but sneak a second look at the young lady in question. She had long black hair and an AC/DC t-shirt, my favorite band.

Pretty much this same group moved around to our other classes together. There was a Thermodynamics course as well as a Calculus class. We all wound up in Physics 101 at the same time too. Chemistry and English 101 were in the afternoon, and Manjeet and I were on our own. In all of these, we took our seats in the back of the room. The Aerospace crowd continued to stare, but no one else seemed interested. After a few weeks, Manjeet and I could begin to tell which of our classmates were on top of things and which ones were lagging a bit. He and I, as well as those of us who were Uplift kids, were usually the most engaged. There was that one girl who had raised her hand that first day, though. She was an unknown quantity. She was always one of those who raised her hand first, and always seemed to listen with a sort of laser focus. The Uplift folks had really already heard a lot of the Aerospace stuff before. At the dinner tables at home, our parents droned on about what they did at work that day. Nevertheless, she was always right there with us in class, regardless of the topic. All I could gather about her was that her name was

Matilda, and later that her last name was Orona. Some old school New Mexicans could recognize that the name was used by the Basque immigrants to Mexico from the Iberian Peninsula. These were converted Jewish people who had fled to Mexico to escape the Spanish Inquisition but had converted to Catholicism anyway. I hope I understand that correctly. Anyway, she was just interesting to me for her intellect, even though all I ever saw of her was when she would look back at us and scowl a bit when we got called on. When we started our study group over lunch about three weeks in, it was exclusively Uplift people, but some of us thought she might be appropriate. We would see her studying independently at the library in the morning now and then. She would see us collecting in our little room and scowl even harder.

Things changed a bit in mid-November, just before Thanksgiving break. Matilda had become an ever more dedicated student and started to get up pretty early in the morning in her dorm room and start running around the campus in the pre- dawn hours. This was not entirely uncommon for the athletes on campus, but engineering students tended to be rather hungover at 6 a.m., so Manjeet and I accepted. We two had decided to go for an early breakfast at the Frontier Restaurant across the street from campus. (Their cinnamon rolls have been famous for about a century). It happened that Matilda was jogging on the campus side of Central just as we rumbled by in our loud yellow monster. Now, it was rather dark, but it seems that our heroine could not help but notice the turbaned profile of Manjeet at the wheel, and also that he had an entirely mundane-looking passenger beside him. She stopped in her tracks and squinted in the dark as we parked and went inside. We didn't notice her at all. This was a shame because she always dressed in tight yoga pants when she ran. (Oops! Too much information!) Then again, we couldn't have afforded to risk wrecking the Maserati from the distraction. We ate our breakfast and then showed up for Numero 9 as usual.

This morning turned out to be unusual, as it happened. Matilda was there at her little study table, already showered and dressed as usual. We ignored her as we had trained ourselves to do, but she suddenly slammed her books down, stood up, and shouted. "Alright, I've had it with you two! Get over here! I want some answers!" The librarian was about as annoyed with this behavior as we were. She demanded we "Take it outside". As no one else had arrived yet for study group, Manjeet and I shrugged and did just that. Manjeet started by registering his indignation with his usual unshakable calm. "We do not take kindly to being shouted at for no reason at this time of the morning," he said, or something like that. She wouldn't have it.

"You two are so weird!" she continued loudly. "You're always together! You sit off on your own, and you go have these meetings before the sun's even up! And now I see you driving around in an awesome yellow sports car at 0 dark thirty! What the hell is the deal! Are you drug dealers or something? Oh yeah, and what's this medical condition you have? Are we all gonna die now, or what?" I started things off incredibly badly when I looked and Manjeet and said something totally sexist like, "She's really cute when she's angry, huh?" which caused her to punch me quite soundly in the shoulder. Manjeet sprang into action as a bodyguard and grabbed her by the wrist, pulling her arm around to her back and restraining her handily. I stopped all this as quickly as I could.

"OK, hold up! I like your enthusiasm, but this is becoming a bit too much like an assault situation for my liking. Manjeet! Let her go, and *you* stop punching people!" I said to each of them. I cringed because I was sounding a little like my dad. Let's see, we were just going to study group, our medical condition is claustrophobia, and my family likes nice cars. Beyond that, if you don't like it, that's your problem."

Matilda wasn't suddenly silenced, unfortunately, and muttered a few more obscenities which I can't recall at present, and neither can she. Anyway, I just kept going. "Look, believe what you want, but our study group has been thinking about asking you to come join us anyway. You are obviously good at this stuff, and you could really help us a lot. So! We're gonna be late if we stand around arguing, and that's against the rules. So come back to our room with us or don't, but we're leaving. So off we went. She didn't follow us at first. But about fifteen minutes later, we saw her through the cloudy glass window next to the study room door. She appeared to be listening or something, so we let her in. She reluctantly decided to join the group when she found out we were pretty serious about doing the work. She never stopped coming to Number 9 until we all graduated. To this day she says she never even thought that Adam Thornhill was really Adam Thorne until I revealed it to her myself. For those first two years at school, we were all just really dedicated students getting really good grades.

As I suppose everyone knows by now if they ever studied my life, my one weakness in life is Math. I had to take a remedial course the summer before I started college just so I could start with Precalculus my freshman year. I think the main problem is that no one could tell me how this stuff had ever come to be discovered or developed or whatever. Sir Isaac Newton just must have had some revelation or something, and Bingo!, now we have calculus. Teachers would just tell me, "Don't think about it, it's a defined process to differentiate an equation." Just do it. When I asked what good it was, it was always, you'll find out later, especially in Engineering. So, I did what they asked, and still, to this day, I can't think in mathematical terms. I have found I just know what I want to accomplish when I design something. I tend to allow the supercomputers to do the math that goes with it. My math grades never matched my other grades. I was never a cum laude guy like my father was as a result. Dad never said anything, but he never got the chance. We'll talk about that later. Anyway, it didn't affect my engineering grades all that much.

At any rate, our first year passed. All of us Nueves really began to appreciate each other. Matilda realized that at least we were more beneficial to each other together than we would have been apart. The one who impressed me most was Manjeet. He was a preposterously no-nonsense sort of student who just bit into the meat of any problem our Profs could dish out. At the freshman level, these problems weren't anywhere near the level we'd be facing as seniors, but it was a trend that has continued throughout the time I've known him since. Never one to show emotion when stressed, he became what the early astronauts would call a " Steely-Eyed Missile Man" when it came to academics. His strengths seemed to be in overall systems design and integration. If a design change were proposed for a particular component or other being discussed, his brain would just click out what effects that change might have on the component overall. Again, this was rudimentary stuff at this point, but his talents in that direction could not escape the notice of any discriminating Engineer. As for Matilda, who by now had graced us with permission to refer to her as Mattie, she was simply the most passionate and motivated person in the group. As we had seen, she was someone who would blurt out whatever thoughts came to mind, which she felt the universe needed to hear. It wasn't that her thoughts were ill-considered or rash either. She just happened to know a lot of things better than most of us and felt that this information had a vital need to be dispensed immediately. This was massively irritating initially, but eventually it wound up just infusing us with mighty doses of motivation. She also was the strongest mathematician in Number 9, which didn't hurt. We later found out that both her parents were killed in a ballooning accident when she was 15. She earned her scholarship as a foster child. She had no siblings either.

Other Nueves that deserve mention and mentioned often they will be. One was Azaan Nazari. He was endlessly fascinated with space propulsion systems, both actual and theoretical. He was an Uplift kid whose parents hailed from Tanzania. Gregor Levinski appeared to be another systems integration type. His folks came to Uplift from Moscow before he was born, and they worked closely with my dad. Juni Sato was knocking their socks off in Computer Engineering and was comprehending the complexities of Artificial Intelligence quite well, judging from what Manjeet and I were hearing from her every evening. She wasn't strictly a Nueve, but she might as well have been. All through this year, I had pretty frequent contact from my father, and Manjeet, Juni, and I drove home every two to three weekends. My father apparently was also keeping tabs on various other UNM students, some of whom we hadn't met yet. He was always on the lookout for new talent for the company and was in close e-mail contact with the faculty to see who might be recruited. Apparently, someone had tipped him off about our Mattie Orona. "What do you think of her?" my dad asked me before the end of the Spring semester of our freshman year. We three were spending the weekend at our house. "I, uh, like her a lot," I said with a more revealing tone than I had intended.

As the end of our first year approached, we whiz kids in Numero Nueve realized that we had all done pretty darn well scholastically. My B in Calculus made me the only person with less than a 4.0. Of course, I realized it probably didn't matter, as my future had never been in question since the day I was born, assuming I didn't reject it for myself. As with all college students, my study buddies began to speculate what they might do to earn a little extra money over the summer. Although I should have gone through the faculty more aggressively than I did, I let it be known that there were several summer internships available out at Uplift Aviation. I didn't disclose how I knew about these, but no one other than Mattie needed to wonder. I had, in fact, conferred with my dad about this, and he had his secretary Frieda e-mail me some applications. She also sent some to the college office, and some students from outside our little group were selected. Of course, all Nueve Mexicanos that applied were chosen, maybe because they would live on-site and save the program a little money. If this was somewhat unethical, it didn't seem to matter, because room and board was all that was offered as payment. I didn't really count because I was already considered an employee of the company.

Mattie seemed especially pleased to get her internship. She actually had turned down a couple of other offers from out of state. Then again, she was New Mexico through and through so Uplift was a priority assignment to her. She was bilingual and often cussed us out in Espanol in our morning meetings. She was an Albuquerquean, first and foremost, a "Burquena" in the local lingo. Burquenas have their own distinctive lingo and outlooks on life. They are also universally beautiful. When Mattie got her acceptance letter e-mailed to her, Manjeet and I informed her that we both had received ours as well. We invited her to come and stay with us for a few days before we were to report to work in T or C. Since she was already due to check out of her dorm room for the summer, she agreed. Knowing that we already had a female roommate, she didn't seem nervous about it. When she arrived there for the first time, she was less intimidated by the above-average level of opulence of the home of two college freshmen than I was expecting. "OK you guys", she said simply. "Which one of you is rich? "Manjeet and I played dumb, or perhaps just were dumb, and each of us pointed to the other. "Fine," she said. "I've been around you two for a few months now. Only an idiot wouldn't know a few Number 9's seem to have some inside knowledge about Uplift Aerospace, even though you try not to let it show. Somebody around here must have connections with some bigwigs down there! Now who is it?" Manjeet and I looked at each other again and shrugged. Knowing she would soon find out anyway, I yelled out, "Juni! Could you come out for a minute and meet Mattie? I think she's on to you!"

Juni was in the midst of packing as well and came out to see who we were talking to. Mattie was very much expected, of course, and we introduced our new temporary roomie. Mattie would be sleeping on the couch tonight, but female visitors to our little place were unheard of at that time, as were all other overnight visitors. "Juni here is the source of all this opulence," I explained to Matty. "Yep," Manjeet said in his usual deadpan. "Her folks are bigwigs down at Uplift. Don't know why they let us stay here. To protect their little girl, I guess." Juni rolled her eyes in the way she usually did when her roommates were being moronic, which fortunately was only when we were awake. Mattie didn't buy in, being a sentient Homo Sapien and all. "And she lets you drive her Maserati for her as well? Wow, pretty generous since she's never with you when you drive it. I'll tell you what, Juni, just let me drive it for you from now on, I'll do better than these jokers any day."

Mattie, of course, had no car of her own, but we didn't doubt that she was probably right. She and Juni seemed to hit it off from that very night. This was fortunate, considering how close we would all be for all the time together still to come that started that evening. We had dinner, did the dishes, and got a good night's sleep before getting up at six to finish our packing. Having only one vehicle, we all would drive down south together. Mattie was duly enthusiastic when we went to the garage and packed our stuff into the kind of smallish trunk. "We're taking this car down to Uplift! Right now?! I guess the Car God answers prayers! You're shitting me, right?"

"What, did you want an Uber or something?" I said. "Wait! I know you don't like Manjeet's driving. I don't suppose you happen to have a license? We know you don't have a car." Matty missed not a beat and flipped open the wallet that she kept in her fanny pack. She threw her driver's license literally into my face. "Hmm, it's current," I observed casually. I asked Manjeet if our insurance would cover her if she drove." "It'd better, considering what we pay for it," he said flatly. "But your father made me promise to do all the driving." "Well, I'm 18 now," I said." And this car is in my name. I guess he's out of luck."

"So are we," Manjeet groaned. "OK Matty, here are the keys. I'll be sitting right next to you. You do what I say, and take it slow, or I'll never let you drive one of our really good cars. "Mattie refrained from fainting, and Juni and Manjeet squeezed themselves into the back seat, looking basically terrified. I clicked the garage door open, and Matty backed out. She began slow and steady, at least till she backed out into the street. Then, of course, she floored it. Being on a rather short side street, however, her racing streak only lasted a few seconds.

"Hmm, it appears your ability to control yourself is kind of lacking," I said. "Tell you what, if you keep to the speed limit till we get to Uplift, we can really do some fast driving. There's a track there we can use if we get permission. Otherwise, no more use of any high-end equipment, and there's a lot of it down there." She really seemed annoyed to hear this, but her brain was throttled for the time being, and she agreed that a drive to T or C in a new luxury sports car was worth some self-restraint after all. It was worth it for her, driving for two hours with everyone on the Interstate drooling at her from the slow lane. Truth be told, I found myself drooling too. Before we knew it, there we were at the Uplift security gate. Mattie noticed that as we pulled up, the guards wanted to just wave us through. I told her to stop anyway. "Sorry, ma'am," I told the Guard on Mattie's side. "You must have us confused with someone else." I forgot about the security sticker on the windshield. "Oh, right", the guard said. I gave my Thornhill name. She rolled her eyes and waved us through. We reported to the cafeteria about on time for our internship check-in program. For appearance's sake, Manjeet and I filled in our paperwork just as everyone else did. Roll was called, and Mattie got her first hint at what was up when several giggles were heard when I answered to "Adam Thornhill." After the formalities were complete, we went back to the car to find our assigned lodging. This time, Manjeet took the wheel. We drove to the apartment complex reserved for Uplift visitors, and there were many of those, and took her luggage up to her assigned room to get her settled in. Interns like us generally got a large studio with a full bathroom and a big study area, plus a kitchenette and a Murphy bed. We told her we'd pick her up for dinnertime at the cafeteria. Most of the luggage we brought was hers, just two modest suitcases and a makeup case. Manjeet and I had enough stuff at home already. She was pretty sure that we were both from Uplift families and would stay at our parent's houses.

"We're headed home now," I said." See you at six."

"Right," she said. We dropped Juni off at her assigned place, then proceeded to dear old Thorne manor. Manjeet had decided to stay with us. His parents had given him permission. He would see them often enough at work anyway, and they would come over to my house to visit him as well. This was a pretty common practice on our frequent weekend visits here, although sometimes we stayed at his parents too.

Dad was of course quite glad to see us when he got home at about six. He wanted to have a barbeque, but I reminded him there would be a dinner provided for all the interns at the corporate cafeteria that evening. "Oh right," he said. "I should go to that!" He always had such a full schedule; he often forgot these little details. Not wanting to be embarrassed, he agreed to stay home while we went to the dinner. But then he remembered something.

"No, I have to, I forgot I'm the keynote speaker."

"Ok Dad," I moaned. "Just act like you've never seen me in your life."

"Oh," he said. "You still haven't told *Her?*"

"Not yet," I said. "I know it can't wait much longer, though. Just trust me, OK?"

"Always," he said.

The cafeteria dinner consisted of roast beef, mashed potatoes, and green beans, along with the drink of our choice. I don't know why I remember that, but it was also the traditional meal for crews who were scheduled for a flight the next day. (This was assuming that the crew member's culture was compatible with such a meal. There were other choices for those that weren't). Dad's secretary Frieda was the Master of Ceremonies. She welcomed us all, Manjeet, Suni, Mattie and I all sat together with a total of about ten interns to a folding cafeteria table. Everything started when my dad showed up and sat with the mentors at the front. After we all got our food, the program started. Dad gave his little welcoming speech and was kind enough not to look at me once during the talk. Frieda then continued with a presentation including things like company history, the goals of Uplift internships. She then displayed a map of the campus on the large viewscreen that had been set up behind the head table. Then she gave a summary of the various departments that were there on the Uplift "Campus." Candidates had requested which departments they had wanted to work in on their applications, but Frieda made it clear that not everyone would be getting their first choice. Juni, for example, had asked for the IT department. I happened to know in advance that she would of course get what she asked for. I had also made a suggestion about Mattie but had no idea what would be decided. Frieda concluded her indoctrination speech by pointing out a table set up on the east side of the room. There were four mentors seated there, each with a fourth of the alphabet printed on a sign in front of the table, A-F, and so forth. We were all to stand up and proceed to the appropriate line and receive our assignment package.

Some of the interns got to work in Dispatch, some in Engineering, some in Design. Others would work in IT with Juni. Still others were sent to the hangers for vehicle prep, and still others with Cargo Management. Only three of us were actually assigned to Flight Crew Operations. Mangeet and my assignments there were known to me in advance. The final selectee was a surprise. Her name was Matilda Orona. Mentors from all the sections were there for the dinner, and after everyone got their packet, they called their department's interns to their own little areas for instructions about what to do tomorrow.

The Flight Crew Ops mentor was none other than Nathan Floatingfeather, my dad's second in command. He nodded at Manjeet and me as we walked over with Mattie. His name was not generally known to the public at large. If it had been, Mattie would have been impressed. Instead, as far as she knew, she was just meeting another Uplift employee. True to character, she immediately complained about her assignment. "I had requested to work in the design department!" she very nearly barked. Nathan had been warned about her directness. "You are Miss, ah, Orona?" he said, knowing full well who she was. "Yes, sir," she said, feigning respect. "I'm not even sure what Flight Ops is," she stated flatly, "but it doesn't sound like a good use of my Engineering background." "Well, Miss Orona," Nathaniel responded. "I must tell you that you were selected for this spot because of all the applicants we examined, you were the only person in the group who appeared qualified for this slot. This is actually the most demanding internship we offer." Manjeet and I managed to suppress our urge to scowl on hearing this. Mattie would be insufferable for the rest of the summer now. "For your information," Nathaniel continued, "Nowhere is a top engineer more vitally needed than in Flight Operations. Nearly every mission has demands that are not entirely met with the equipment available on a Shrike Spacecraft or its supporting infrastructure. Equipment must be refined or completely redesigned in accordance with crew requirements on a damned-near daily basis." He looked at Mattie intently. "I hear you are an immensely energetic person. You're going to need to be, believe me. You three and I will be spending a lot of time together."

"So why are these two here?" Mattie asked. She seemed dead serious. "They're kids of some big wigs or other, aren't they? Isn't that right, Mr. ThornHILL?"

Nathaniel stayed solid." These two are here because their grades very nearly match your own," he said. "Plus, the University knows about their achievements in leadership of that little study group of yours."

"And because my parents are bigwigs," Manjeet piped in.

"Never mind", Nathaniel interrupted. "Adam, I'll need your little team here fed and equipped by 8 a.m. tomorrow. Report to the flight briefing room. Oh wait, I forgot to have you all sign those consent forms and those waivers of liability as well. You're all over 18, as I recall." The need for everyone to sign these forms had been announced in the induction dinner already, so we were not surprised. We all signed. Only Mattie failed to suspect that she was signing on for much more than she anticipated.

Chapter Five
Taking Wing

From the personal recollections of Nathaniel Floatingfeather…

I am taking over the narration of Adam's time as an intern at Uplift because Adam's transmitted diary entries are really a bit dry, considering how memorable these times really were for those on the outside looking in. Anyway, Mattie, Manjeet, and Adam arrived exactly as scheduled at 8 a.m. at the Flight briefing room. They had had a full breakfast at the cafeteria. Fortunately, they had all had the appropriate dinner the night before. The briefing room had a large conference table, a projection screen, and a chalk assignment board on a side wall. This had been filled in with four flight numbers, the spacecraft number or the flight, the flight crews, and the mission objectives, along with the expected flight duration. There was a side bar with coffee and refreshments as well. When my three interns arrived, I was standing at the podium at the head of the table. All twelve seats around the table were filled and another twelve or so seats were positioned at the periphery as well, with three exceptions. I had the newcomers sit in those. These meetings were also recorded. I'll tell a lot of this story in first person.

"Welcome to morning briefing," I said. Often, this briefing was chaired by one of my flight directors. Today was a bit different though, due to the special nature of one of the flights. I began listing the four flights in accordance with the expected departure times. Flights were generally coded 1 through whatever, in this case, 4 for four flights, followed by the date. So, these were flights 16280 through 46280, the day being June second, 2080. Flight One was a three-person flight to supply the International Space Station with food and supplies. Two was a satellite retrieval mission with the transport of waste plastic to a rendezvous with the sun. Four was a delivery flight to the Uplift Medical Research Orbital facility. Mission Three was the one of most interest to me. It was a two-day orbital research mission with a fairly noteworthy crew. As the crews began to listen, three of the mission crews were already familiar with their assignments for the day. The One and Two crews merely took notes to confirm expected flight departure time, flight paths and orbital requirements, fuel consumption expectations, communications channels, and other data that may or may not have been standard for their types of flight. All this information was already preloaded into the computers of their mission assigned Shrikes. The pilots and flight Engineers really didn't do much flying themselves. These features were all automated. But humans were needed if things happened to deviate from what was expected. Most crews flying with Uplift already had vast experience. Everything about this meeting was actually quite mundane and routine. Or at least they were until Mattie Orona glanced up at the assignment board. Again, all these meetings were recorded and placed in Uplift archives.

On this day's meeting log, I recorded a moderately loud "What the hell!" during the Flight One briefing. Adam and Manjeet were seated on either side of Mattie, the one who broke the usual expected silence. "Shush" and "Just relax" were whispered by her two companions. She didn't say anything further until she was called upon. The agenda progressed to Flight Three in due Time. "Adam," I began. "You have already submitted your flight plan, I see. Take off will be at 1400 using runway two. Winds are from the northeast at 4 miles per hour. Visibility clear. Course is southeast over Texas with orbital insertion beginning at 40,000 feet over Barbados. Orbital insertion at 12 miles altitude with reinsertion burn at T plus 48 hours over Siberia. Manual control should commence at T plus 48:30. You've practiced this for a couple of months now in the simulator, I see. "That's affirm," Adam said. This is when Mattie lost just a bit of what was left of her composure. "You've got to be kidding me!" she all but yelled. "You're piloting this flight! This Spaceflight! There's no way I signed up for this!"

"Actually, you did," Adam turned to her and said. "Just last night, you signed a waiver, remember?"

"No one mentioned anything about going into orbit today, or did I miss something?" Mattie gasped.

"Well, I thought the words "Spaceflight Operations" might have been kind of a clue," Adam replied. "Besides, you said rides on faster family vehicles sounded kind of appealing." Mattie looked again at the assignment board and saw the name Adam Thorn written in the pilot's slot. The copilot was to be someone named Ana Koselka, and Manjeet was slated as Flight Engineer. Her name was in a "Crewmember" slot. Mattie kept going. "You mean you've been a freaking Astronaut all this time I've known you, and you never let on? I kind of figured you were really a Thorn for a while now, but don't you think you should have told us from the outset? How many missions have you commanded anyway?"

"Well," Adam started slowly. Dad's been letting me go on missions for about four years, I think. This is my first time as lead pilot, though". Mattie sort of moaned.

"Relax there, crewman," the copilot barked. "As you know, these systems are pretty much computer controlled. Besides, I've flown about thirty missions since I started here, and ten of those as a pilot. If I trust Adam to do the flying, then there's no reason whatsoever why you shouldn't. But if you want to stay down here in the gravity field, we can go get damned near any other intern to take your spot".

"Oh, hell no," Mattie said firmly. "You guys just wanted to see what I'd do, didn't you? Well, guess what, I've dreamed about doing something like this for most of my life. It's just that it's not something you just announce at the last second. We'll talk about this later, Adam! I'll be needing you to treat me way better than this in the future, by the way."

"Duly noted, crewman," Adam answered a little too flatly.

"Alright everyone," I announced from the podium. "If you're not already suited up, report to crew prep and start denitrogenating. Sorry about lunch, crew three, you can eat in orbit. Adam, please brief Ms. Orona on your flight plan. Let's stay focused, people. This is the last day we need any screw ups, especially since it would also be the first time we would ever screw up!" With that final pronouncement, the crews broke into their own groups, and Adam led his small enclave to the Crew Prep area, formally called the Ready Room. As I didn't want to miss this, I followed along behind.

There was a short walk outside the Operations building to the Main Hanger facility, more towards the middle of Runway Number One. We entered through the main hanger door, where the day's four Shrikes were being fueled and given their pre-flight check-outs. The place was swarming with activity. There was the mist of supercooled Oxygen and Hydrogen being topped off, although the main fueling had been done before the ships were brought inside. Adam led his three crewmates to the predictable "Crew Ready Room" sign and punched his security code to gain entrance. There was a moderate-sized lobby that led to several side hallways. Adam pointed to the hall marked "Suit Prep Area" and led his crew that way. On each side of the hallway, there was a door marked with the same Male and Female logos found on most public bathroom doors, one on either side. As this crew had two of each gender, the group split in half. Ana signaled

Mattie to follow her in. "This gender thing isn't really strictly enforced," Ana told her. There are crews who work together pretty routinely, and they usually all suit up in the same room. This crew is mostly semi-newbies though, so we'll separate for now. Kind of silly since we'll all be locked up in a fairly tight spacecraft for the next couple of days."

"Whatever," Mattie said. "Those two have probably never seen females dressing in their lives, from what I can tell, so why distract them when they really need to focus on their jobs?"

"Good point," Ana replied, quite seriously." By the way, now's a good time for a potty break, whether you need it or not."

The male and female areas were just mirror images of each other. There were six reclining lounge chairs covered in vinyl fabric. Each group was met with a suit specialist who took their pulses, blood pressure, and temperatures before stepping with them into the suit storage unit. These folks had lots of practice estimating suit sizes. More tenured crew members had specific suits assigned to them though, so only Ana had one of her own. The other three would too, eventually. Adam had taken a few flights, but not enough for a permanent spacesuit assignment. Suits for Uplift were quite simple, rather like the ones used at Space X and other commercial space enterprises. As there would be no space walking on this mission, the suits just resembled stylized rubber jumpsuits with helmet labeled "Uplift" and serial numbers at the base of the neck area. Everything was navy blue. Shrikes still had pure oxygen crew compartment atmospherics, so pre-breathing from portable oxygen units was necessary to purge nitrogen from the bloodstream before boarding. This was unchanged from the days of early spaceflight and kept you from getting the bends in the lower pressures experienced on board. Proper oxygen saturation of the red blood cells was maintained, and you could get out of your suit as soon as it was confirmed that the Shrike's cabin was airtight after attaining orbit. Nevertheless, one had to cool one's heels beforehand for about three hours while this nitrogen purging took place. Fortunately, the helmets had com systems to allow communication between all crew members, as well as with the world outside your helmet. These exchanges were recorded and stored in the Uplift Cloud.

"OK, so what's this mission about?" Mattie asked Adam when the crew was reunited in the ready room.

"We're just taking some animal specimens into zero-G to see how they respond," he answered.

"Oh great," Mattie groaned. "They've been doing that for years on the Space Station, decades even. Is this to show to the Kindergarteners of the world? What are they, mice or spiders?"

"I think I'll surprise you," Adam said. "It's just a little project of mine I designed to satisfy my curiosity. It's kind of been done before, but not on this scale. What's different is the containment system. I've been working with the design team to build it. We rich kids have weird hobbies."

"Like making space cages for hamsters? Yeah, that's kind of weird," Mattie zipped back.

"Look, don't worry about it. Your job will be taking notes on how it all is working." Adam explained.

"I think I'm feeling like one of the hamsters," Mattie pouted. Adam cut her off and began speaking in flight jargon with the other three for the remainder of the prep period.

The time went rather quickly, at least for those members of the crew not named Mattie. She seemed to need to calm herself, closing her eyes and taking deep breaths. She would have read a magazine if she were used to reading through the glass faceplate on her helmet. She did seem to recognize some of the technical aspects of her crewmate's conversation, adding an "Oh, right" or "I hadn't thought about that" from time to time. She was definitely paying attention. At about 12:45 p.m., I informed the four of them that they had been cleared to board their Shrike, number 337. This information was not really too relevant, as all Shrikes were pretty much were the same, but we managers need to record such things. There were so many Shrikes and so many missions that some crews had taken to calling the hanger "The Truckstop." It really did bear some resemblance to one. Being the only crewmember who had never been around the Uplift flight line, Mattie was clearly in awe of the whole mess.

One enters a Shrike from a square hatch on the pilot's side of the cockpit, just behind the pilot's chair. There is also an access hatch in the ceiling if one docks with an orbital structure. On the ground, you go up a ramp sort of like small airplanes have in small airports. You ingress through the side hatch into the cockpit. This is not too hard with the thin spacesuits and the little hand-held oxygen concentrator you're carrying. Adam went in first, as is customary for the pilot. Mattie was next, and Adam helped her get situated in her right rear cockpit seat, then switch her O2 line to the spacecraft port. He then did the same for her suit power and communications line, which was more of a USB cable. Her seat also had a fixed laptop-type screen in front of it, which was turned on. Adam then got situated at the Pilot's seat position, in front on the left, with a very large touchscreen control console similar to most contemporary spacecraft. Ana entered next at the copilot's seat on the right. Manjeet was last and took the back left seat, the Engineer's position. This had a fairly large flat screen in front of his seat as well. The screen displayed numerous virtual dials and readouts which looked not unlike something one would see on an old predigital age flight engineer's console, only with a blue background, not a metal one painted black. Mattie noticed that the copilot's station was set to display navigational-type information on the right side of the screen. She heard Adam and Ana begin their pre-flight checklist tasks, which sounded exactly as one would expect. She couldn't hear any sounds that weren't originating from outside her helmet communications. If she had, she would have heard the tow vehicle backing up to the front of the Shrike to latch on. She did feel the spacecraft suddenly lurch forward as it began to tow them to the launch runway at about 13:45 p.m. She began to get a bit nervous.

None of her companions seemed to share her nervousness. The technical chatter continued, of course. Manjeet piped in occasionally with fuel tank pressure and temperature reports. The control tower chirped in with weather and wind reports and the like. Such normality was reassuring to Mattie, along with the knowledge that Orbital Spaceflight from this runway was about as normal as any commercial flight in a conventional airport. The Shrike itself was nearly as long, but much wider and thicker than a large passenger jet. In fact, two twin jet engines were being raised (deployed) from the upper surface of the shrike, just in front of and on either side of the vertical tail. These were just now being powered up for takeoff. She felt the vibration but heard nothing but the technobabble in her earphones. Then came the announcement. It was precisely 1400 hours. "Flight 36280, you are cleared for takeoff. Godspeed and good luck!" The voice was that of Frances Thorne.

Mattie could feel the vibration that seemed no different than any jet aircraft powering up for takeoff on any runway in the world. This is because it was no different. Then came the release of the brakes which sent the craft, now an aircraft, lurching forward. This plane, though heavier than most of its size, had plenty of thrust behind it and pushed its passengers back into their seats with more than everyday force. The craft was a lifting body, in other words, contoured to just jump off the ground as soon as its aerodynamics would allow. The take-off angle was fairly steep, about 75 degrees. Mattie liked it more than she liked the Maserati. The beast kept climbing. It shot like an arrow across the high desert of New Mexico, rising and rising over the Texas border until the prairie lands gave way to the Caribbean Sea. There were no windows in the back cockpit for Mattie or Manjeet to look out, but Mattie noticed an option for "Exterior View" on her touchscreen, so she raised her oddly heavy hand to select that option. The screen showed a bright blue sea far below, with occasional island-like landmasses dotting the view. A small altitude reading at the bottom of the screen showed 39,970 feet. Mission elapsed time was already 49 minutes, 20 seconds. "How did the time pass so quickly?" Mattie thought to herself. She felt more slight vibrations as the outboard jet engines were retracted back into the body of the Shrike. The engine hatches closed to seal the exterior to its natural smooth contour. The crew was mostly weightless now, and the ship pitched back to a near vertical position, the nose shifted toward the preselected course. A countdown was broadcast through the earphones for "Main Rocket Engine Start. Five, Four, Three, Two, ONE, Ignition!" Then, "Pow!" for lack of a better word. Mattie had never felt anything like it. But hell, whoever had who hadn't ever flown to space, especially orbital space. It took about three minutes of being pasted to the back of your seat, unable to move anything but your eyeballs and fingertips, "Eyeballs In," astronauts call it, as your eyes get plastered against the back of the sockets. The force in a Shrike is about 3.5 times the force of gravity and lasts nearly 4 minutes. Then comes the engine cut-off, followed by a few small course correction burns. After that, a sudden stop of the force on your chest, followed by complete weightlessness. No wonder Ana suggested the potty break that morning.

In point of fact, both the males and the females on all Uplift flights wore briefs under their suits. No records are kept regarding this particular area of flight activity. However, the conversations that will now be put into narrative form are compiled from audio recordings taken on all Shrike missions. These are transmitted down to mission control as part of the "Black Box" record to be used in case of mishap or of legal problems. So now that the disclaimers are complete, we can report that after the orbital insertion shutdown, the four members of flight 36280 began to release their restraint belts and begin floating in the control cabin. Mattie alone reported a touch of nausea, but whether by design or pure chance, the crew had not eaten lunch before launching today. She alone was also the only crew member who had not experienced weightlessness, but she didn't complain at that time about the disorienting effects of not feeling any sense of up or down. Manjeet called out the shut-down checklist as various systems were activated, and others deactivated. This was a purely orbital flight. No rendezvous were planned or anticipated. As with all spacecraft operating in the solar system, care was always taken to avoid having only one part of the spacecraft sit too long either in the sun or in the shade. The automated system for attitude control was activated in order to begin the familiar "Passive Thermal Control," or PTC roll, to help ensure a more even distribution of heat. Astronauts had been doing this when reaching orbit since the very early days. Shrikes had their own little external thrusters to control these relatively minute ballerina movements by the spacecraft. Another part of the checklist was a cabin pressure check to be sure there were no air leaks in the cabin, heating and cooling fluid checks, battery power and generator checks, and fuel status verification, along with several other checks. These were generally done automatically but monitored by Manjeet. Adam and Ana were checking on Navigation accuracy and communications systems. All of this took about an hour, and when all was completed, Adam was able to pronounce, "Let's get out of these suits!"

He led the way through a hatch space which he opened by the throw of a switch at the control console. This led to the lower deck, which held the crew quarters, galley, and access to the hold behind the back crew compartment bulkhead. This also emulated the old Space Shuttle design, except the airlock to the cargo space was on the far side of the bulkhead, as was the rear viewing window and robot-arm controls and so forth. On this mission, the cargo door would not need to be opened, and the rear compartment was sealed and pressurized with oxygen. No EVA suits were aboard. Adam took it upon himself to give his newbie crewmember a guided tour of the crew quarters compartment, as any good captain would. Though not luxurious by any stretch of the imagination, the accommodations were quite comfortable compared to those provided to the crews of the old Space Shuttle flights. As in the International Space Station, each member was given a cylindrical, stall-like "bedroom" placed vertically in relation to the rest of the cabin. There were four of these, each at a different corner of the lower deck. For missions with more than four crew members, the other people could sling hammocks in the spaces between the cabins or take one of the cockpit seats upstairs if they preferred. Crews did their best to keep the sleep periods during nighttime hours back at Uplift, also emulating NASA practices. When all the Hammocks were stowed, the center space acted as a galley and meeting or workspace. The center front bulkhead also had a toilet and bathing space. The toilet was the tried-and-true suction seat type, with relief tubes for the males. Showers were really just moist towelettes with a container to dispose of them after use. The place was just a recreational vehicle that could be used in zero-G. Mattie took all this in attentively and seemed to be having a really good time in the process. Adam pointed out the back center bulkhead, which held the egress hatch for going through to the cargo bay. "We'll be showing you our mission hardware back there in a minute," he said. "Here, I'll show you your hotel room now." He floated her over to the forward right cylindrical space, which had a folding rigid curtain that could be closed for privacy. A vertical sleeping bag was slung from top to bottom, with a zipper and opening at the "top" for her face to be exposed to the open air. Near at hand were heat and airflow controls, a reading lamp on a flexible rod mount, and other lighting controls. There was also a USB port. Wi-Fi was provided conveniently. On the opposite side of the sleeping bag was a laptop on a movable mount. It currently displayed the same information shown on the screen in front of her launch seat station. Mattie tried to imagine sleeping in this box tonight.

"Ok, everyone!" Adam shouted. "As soon as we're all settled, our mission schedule gives us 45 minutes for lunch. I'll start breaking out the food packets. Since no one had much in the way of personal effects to secure, apart from their spacesuits, this didn't take long. A standard fare of edibles was included for the crew, as this mission hadn't been planned in advance. No requests for dietary preferences had been solicited before the flight. But there was already an adequate variety to satisfy most crews. As many crewmembers originated from India in Uplift, Manjeet was not disappointed. He expected that there would be an adequate supply of satisfactory food aboard that would be to his liking, and there was. All manner of beverages were stocked in the refrigeration unit. Lunch consisted mainly of sandwiches. A type of non-crumbly bread had been developed for flights that didn't fill the air with particles when eaten. Four seats were arranged around the round table in the center of the common space which were equipped with straps that allowed one's bottom to stay snugly in the seat. Adam distributed the edibles and then strapped himself in beside his crew. Food and drink pouches also had Velcro on the bottom to keep them from floating away from their owners. Discussion about the afternoon's activities started as soon as everyone started chewing.

"So, this mission has something to do with animals?" Mattie asked. "What's the point? Do you want to build an ark in orbit or something?"

"Well, who knows?" Adam replied. "Maybe I just wanted to see how well you could do as a member of a flight team. You've intrigued me for months, you know. You might have figured out by now that I'm quite likely to have some management responsibility in this company at some point. I'm interested in finding people I can work with whenever that time arrives. I happen to think you have a lot of the talents I've been looking for. I must admit, though, that I've been worried that you have some problems with aggressive behavior. I'm wondering how you'll take to being a member of a team. The best teams need to have leaders, but those leaders need to be calm and respectful to all the other members. At least they do here at Uplift. I thought I'd put you under a little stress, to see if you can calm down and do the tasks assigned to you without making everyone else miserable."

"Oh, thank you so much!" Mattie grumbled. "You shoot me out into orbit with no warning and then tell me I'm responsible for not screwing it up! Well, I've got...."

Adam's face became stern. He cleared his throat rather loudly. Mattie realized what she was saying and shut her mouth soundly.

"Oh, I guess I see your point! Am I that predictable? Man! Here I am, getting a chance to do what I've dreamed of doing my whole life, and I'm screwing it up! I guess I've just been on my own for so long, I never got a chance to learn how to behave in public. But really, if you're saying you just made a space flight solely to test my interaction skills, that is a little hard to swallow."

"No, that's only part of it," Adam said. "There's also the part about completing a task under stressful circumstances and learning what resources you have. Now as for the job at hand, you recall that it has something to do with animals?"

"Uh huh..." Mattie hummed.

"Well, this is a little something I've been bantering over with my father for the past year or so," Adam replied. "I've gotten kind of obsessed about the idea of maintaining a permanent presence in the space environment. Not just like on the Space Station where people and supplies come and go, but where you can live out your life up here without ever needing to be resupplied from the ground or go back down when you're homesick."

"You sound like some sort of doomsday fanatic!" Mattie said. "You'd better tell me we're still going back down tomorrow!"

"See, that's what I'm worried about!" Adam said. "I love that you speak your mind and all, but I'm hoping we can get you to make calm and rational arguments occasionally rather than just shouting at everyone. How about something like 'How will this mission contribute to obtaining the stated goal?'"

"Yeah yeah, OK." Mattie groaned. "What you said..."

"Better, thanks." Adam continued. "In the case that permanent residence becomes feasible, one vital part of the equation would be maintaining a food supply. There are many orders of magnitude of other problems, but today we're talking food. Now, obviously, we've proven that food crops can be raised in zero-G using hydroponics and so forth, but what about animal protein sources? Poultry maybe? Who knows how that would work out? Chickens in space! No? I hate the idea of eating rodents too, don't you? Would you believe I even asked my dad if we could put a cow in the cargo bay to see how it coped? He put the kibosh on that one! Then I had an epiphany. I had him build a containment vessel and put it aboard this Shrike. Our next job will be to go to the cargo bay and see how our critters did. Now, let's just do the dishes and we can get started." There was the small matter of collecting up the disposable trash. Ana and Manjeet requested a moment to go brush their teeth and use the restroom, but soon everyone was ready.

Indicators near the rear hatch showed that the cargo bay shared the same pressure as the crew deck, and that no air leakage was detected on the other side of the hatch. It was also possible to turn on and off the cargo bay lights, but these were already on. Bay heaters were also already on and showed the ambient temperature to be 68 degrees.

"Looks like we are Go for cargo bay ingress," Manjeet announced. The door was locked and sealed electronically. He punched in the access code at the panel by the hatch. Access was granted, and the hatch swung inwards. One by one, the crew floated into the space beyond, entering the cargo bay. The bay was spacious when empty, measuring a healthy 100 meters in length, 20 in width, and 15 in height. This bay was not empty, however. Approximately 5 meters in from the hatch, there appeared to be a rather thick clear plexiglass bulkhead. There was a space between the lateral bay bulkheads and what was evidently a large square plexiglass cube or box. Looking up, it was seen that the box also had a bottom and a top. It dawned on everyone (except Adam, who obviously knew already) that this box, which didn't extend to the other end of the bay but maybe only five meters back, was full of water. There were air bubbles circulating through the water, although they didn't flow upwards as they would on Earth, but rather in random patterns. They moved slowly as well, like fluid would if spilled in the zero-G air outside. A pump could be heard which must have been circulating the fluid contents of the tank. And moving about in this fluid were what appeared to be several dozen...

"Fish?" Mattie gasped. Ana and Manjeet did the same, having not been previously briefed about anything other than flight-related data for the mission. "You're doing this because you want to be served seafood on your orbiting space station?

"Calm down, now. Remember to be objective in your observations." Adam reminded her. "Similar things have been done on a smaller scale before. I don't think it's ever been done with the objective of having a renewable food source on a zero-G station though. Now, you might have noticed we used a bit more than the usual amount of fuel to attain orbit today, did you not?" Adam asked the flight crew trio.

"True enough," Ana admitted. "I hadn't been briefed on the weight of the cargo, but there's always a lot of variation on an Uplift flight. You and your dad were OK with it, so I didn't question it."

"Well, just for formality's sake," Adam said. "I feel pretty sure you can start questioning it from here on in. You'll be commanding your own missions in the very near future. Mission safety demands you speak your mind, so long as it's constructive," he smiled, looking at Mattie as he said so. "Sooo, if animal protein is needed to feed future long-term space crews, and orbital cows and chickens would be too hard to manage..."

"Why not use fish?! "Mattie blurted out. "Of course! Fish are already zero-G critters! If you maintain a clean aquarium and feed them, you can have a perpetual supply! Damn, Adam! You're smarter than I gave you credit for!"

"And if we fit the tanks out with some algae and water plants, we can generate a little oxygen of our own, I believe. We'll figure that out some other time. What we're doing now is watching to see what these fishies think about being in orbit. We'll go another day and a half and then ask them their opinions. As of now, however, I don't see any dead fish floating around. There's plenty of fish food suspended around for them to eat, and the aerators appear to be functioning. Look, we have fully grown sea salmon, tuna, and halibut all swimming in front of us. Too bad we can't cook them up this afternoon. Now, Mattie, as a mission specialist, I've sent the flight recorder data to your laptop. There should be a visual feed from that camera up there (Adam pointed to a small camera located above and to the left of the exit hatch) that should show us how the specimens here reacted to an orbital launch. Just be grateful I didn't have a bunch of electrodes inserted in their little fannies."

"Any reason we need to stay back here?" Ana asked.

"No, anything else we need to do can be done in the crew compartment on our laptops. The tank is running on automatic. The lights back here will dim on a timer to emulate ocean conditions and the Earth's rotation. After you, folks." Manjeet took the lead in reopening the hatch and leading everyone out to the lower crew deck. Adam went to his bed compartment and detached his laptop from its bulkhead stand. He went to the center dining table, strapped himself into the chair and stuck down the computer to the Velcro provided. Everyone else did the same. He asked everyone to click on the "Flight Mission" icon. A variety of new icons appeared. "Let's check on Cargo Bay Video." Everyone did as he instructed. "OK Mattie, you're the Mission Specialist." Adam continued. We'll all see the same thing regardless of who clicks on it. Let's see what was going on in the cargo bay when we lifted off". This wound up being fairly interesting footage, especially if one found fish under duress to be entertaining. The tank was completely filled and had no air pockets. There was therefore no sloshing around of water in the tank. There was quite a lot of sloshing around of the fish themselves though. The aerator kept the water inside in constant motion, as seen by the bubble patterns. When the Shrike first started down the runway, the fish moved almost in unison towards the back of the tank. It must have seemed like swimming into a current or a tide, and they all seemed challenged to swim against it. This was only the case when the ship was accelerating. At constant speeds, everything returned to normal. This is true of any passenger in any moving vehicle. What was really interesting was what happened when the rocket engines were ignited. The fish were faced with a near tsunami of acceleration, and all were suddenly flattened against the rear glass of the tank, all piled one on top of the other. Since the water was being pumped around the tank and, therefore, over their gills, there was no more stress on them than there would be in a brief period of brisk current that they would experience in an ocean storm. When orbit was reached, everything calmed down again. They were none the wiser and went back to swimming as usual.

"Well, that was exciting," Mattie said, returning to her usual demeanor.

"Actually, it was pretty boring," Adam responded. "But in this case, boring's what we were looking for. I'm betting that a zero-G buoyancy in orbit is the same as it would be on Earth, and these guys could care less either way. So, Mattie, if you could check on the fish every hour or so, as well as the water chemistry, temperature, and pump efficiency, that would be great. See that "Tank Status" icon? It will record the data for you. Just let me know if anything changes drastically, or if any of the fish die. That's the life of an engineer! I think all of us should go back to the flight deck and see how the ship is doing. Dinner will be at 1800.”

“I hope we’re not having Fish!” Manjeet said with his monotone joviality. Everyone scurried off to their assigned tasks. Mattie strapped in at the crew quarters table and went to YouTube on her laptop. The banter on the ship changed to chit-chat with a bit of technobabble heavily mixed in. Manjeet lost his super serious professional demeanor quickly. Mattie had always known that he and Adam were really, really close friends. Manjeet’s favorite topic this evening appeared to be their friend Juni. Her female intuition told her that Manjeet and Juni must be an item in some way. She’d have to look into that, but for now, she moved on. She looked around. There were so many things for the crew to do on an orbital spacecraft, it was more dizzying than vertigo. Such a frantic pace! Every system needs periodic monitoring, and radio chatter from mission control is as frequent from Uplift as it was and remains from NASA. It took over an hour for the required protocols to be completed. Mattie did her own particular part of this duty and watched the fish swim on her laptop. No problems occurred with the tank. About 1730, the crew regrouped in the mess area, which Mattie had never left. Adam announced that the dinner for the evening was vacuum-packed turkey and peas in gravy, along with a protein bar. Everyone chose their own beverage.

“We contract our food packs with the same guys that make MREs for the military,” Adam announced. That’s a pretty big company. We do at least four flights a day with at least four crew members per flight. Usually only take Thanksgiving and Christmas off. That’s a lot of food!” Adam chattered as he placed the packs in the microwave.

“You’re pretty on top of things for a college student commanding his first orbital mission!” Ana observed.

"Well, I've been around my dad my entire life!" Adam shot back. "Uplift is pretty literally my family. How's the fish tank, Mattie? "

"All fish systems go!" she giggled. "So, did you design this tank or what? After looking at it pretty closely, it occurs to me that it's really pretty elegant."

"I'm sorry to say that I just asked my dad if someone could throw it together before we got down here," Adam admitted. "I did tell him the parameters I wanted, and what sensors and so forth I thought would be helpful. I did make arrangements with the fishermen myself, though."

"I think I'll make it a goal of mine to work in your design department," Mattie said. "The Spacecraft design part, though, not aquariums."

"Yeah, we already have aquariums staffed, no openings that I know of," Adam said, with a few accompanying chuckles from the others. Dinner progressed amicably. The table and galley area were tidied up, and the schedule for the rest of the day was marked as a "Crew Rest Period." Ana was slated to spend the night in the cockpit, sleeping in the Commander's couch with earphones on in case anything untoward occurred. She said goodnight and popped upstairs. Manjeet was obviously very tired and excused himself to go zip up in his bedroom. Mattie and Adam were left at the table. If they weren't already floating there, they would have put their feet up. It was only about 1815 pm.

Mattie couldn't help but be pretty informal with Adam. They had known each other since meeting uncomfortably back the previous fall. They were friends, although distant friends, but then all this happened. "So why did you bring me up here when you obviously don't like me that much?" she asked him.

"I have to keep everybody at a distance when I'm away from here," he answered. "I'm a billionaire's son. You never know when somebody's gonna try to manipulate you, or kidnap you, or whatever. That's what my dad says, anyway. Plus, I'd just like to get through school without being bothered. I've got no time for socializing or drinking, or having wild love affairs, or any of that. I have to succeed at school, that's for sure. My Dad seems to have foreordained that I can help him run this company. But that's fine since I'm the one who asked him if I could."

"Did you ask him to bring me down here? I need to know," she asked. He hesitated a second or two.

"Yes, I did," he responded slowly. "Don't think it's for any reason other than you seemed the best candidate I could think of. I hear your passion for this stuff every day in study group. I see you in class. Look, I'll be needing great people to work with as I get more into all this. I don't think running applications through a computer screening program will quite do it for me. I need to see people for myself. I have to find actual people, not applications. Real people. I didn't know at first who it would be, but it happened to be you. You just fit, that's all. You're not the first. I met Manjeet when I was six years old. Juni too. They want to stick with me, and I'll stick with them. Whatever happens."

"Like flying fish into orbit?" Mattie giggled.

"Oh that, and a lifetime's worth of even weirder shit".

"You've got plans, eh?

"I hesitate to tell you, but you can find out if you stay with me, and if you get a lot more patient."

"Yeah, patience will help, you've got that right. Any other plans you care to bestow upon my humble person?"

"Um, yeah. It appears that Manjeet and Juni are requesting to share the same room back in Albuquerque. He requests that no one tell his parents about this. Sikhs really prefer that their sons don't marry non-Sikhs. She's planning to convert, though. It doesn't matter if she's not of Indian descent, that's allowed. Manjeet has to ease them into the idea, but that will take time. Meanwhile, what they don't know won't hurt them. Unfortunately, we'll have an empty bedroom at the house."

"Are you asking me to be your roommate?"

"There you see? You do have a wonderfully analytical mind. And once again, you're the best candidate for the position."

"It's not because you have the hots for me then, right?

"I'm not sure what the hots would feel like if I had them. I've lived a sheltered life, you see. Maybe we should run a thermodynamics analysis later on. It's just possible I'm exceeding tolerable limits."

"I'll take it under consideration, like a good engineer," Mattie said. "Meanwhile, I guess it's time to see if I can sleep up here. Goodnight, Mr. Thornhill."

"That's Captain Thornhill, for the time being anyway. That's where the patience part comes in." Mattie floated off and zipped up the door to her little cylinder bedroom. The light went out after about five minutes.

After the prescribed eight hours of "crew rest," the four co-flyers were up and bathing, shaving, and the other unmentionable parts of being biological. Breakfast consisted of desiccated bacon, instant oatmeal, and an analog of a former NASA staple they used to call Tang, or instant orange juice. Mattie checked on the more aquatically-inclined crew members and announced that all was well. "All fish present and among the living!" Mattie announced.

"And the tank has held up?" Manjeet inquired.

"Yes, thank heavens," Mattie said, looking for verification on her laptop screen." Can you imagine if it hadn't?"

"I assume the cargo bay would contain the spillage," Ana said. "You know, I'll bet this flight was as much about testing the tank as it was about entertaining the fish."

"You'd be right there," Adam asserted. "I don't think putting that much liquid water into orbit at once has ever been done before, not even for the Space Station. I think it will be a nice precedent in the future. You'll have to be a real space geek to know about it though. We're not going to broadcast it to the world or anything. Just a trade secret, you know?"

"Damn, there goes my NBC special report!" Mattie joked. "And no "Space Fish" miniseries either!"

"Right again!" Adam said. But it was time to play captain once more. "Well, we need to keep an eye on the aquarium for about another four hours. That's you, Mattie. The rest of us need to do some housekeeping and some pre-landing checklist items. Our deorbit burn occurs at 11:30 today. I forget the T-plus number. We'll be reviewing the deorbit procedure this morning, of course. Let's get on it! The three flight crew members proceeded to float up to the cockpit like the good space farers they were, leaving Mattie to clean up breakfast leavings and tidy up the galley. Adam glanced at her as he left, thinking that Mattie's flowing black hair floating around her head in weightlessness really was pretty darned hot after all. The few remaining hours passed rapidly, and Mattie found herself making her final entry in the Mission Specialist log. At about 10:45, the topside crew came back below for a quick bit of lunch and a bathroom break. (These things were important!) Adam had everyone secure their belongings, zip up their rooms, and redon their spacesuits. Then they all went topside and strapped in for reentry. Automated sequencing fired the rection control rockets and turned the ship 180 degrees to assume the proper attitude so the big rockets in the back could fire against the forward motion that was keeping them in orbit. "You know, it really doesn't matter what happens to the fish now," Manjeet said. We were just paid to bring them up!"

"Good point," Adam said. "I can sell you some when we land. Oops. Get ready, everyone!"

At 11:30 precisely, the spike jet engine in the back lit up, pushing everyone back into their seats. After a fair spell of weightlessness, it felt much more definitive than when they lifted off, even though the actual force was a bit less. The force was eyeballs in again. When the engines finished firing, the spacecraft again realigned itself so that the nose was now at the front in relation to the motion of the ship itself, and the nose was pointed slightly upward. This allowed the belly of the Shrike to begin to absorb the heat of reentry. The whole belly was one large heat shield, one piece except the landing gear covers the flaps and ailerons, and the bottoms of the rear horizontal stabilizers. The craft continued to slow at a body crushing rate, with an ionized cloud of vapor forming around the ship that was visible through the cockpit windows. As with all such craft during reentry, no communication with the outside was possible during this phase. Also as with the old space shuttles, the whole experience just wasn't comfortable. Once the ship descended into the thicker part of the atmosphere, however, the Shrike spacecraft proved far superior to the older craft. Once slowed sufficiently with the flaps, those ingenious retractable twin jet engines were extended just beside and slightly in front of the vertical tail fin and then activated. The ship executed the same S-shape course the shuttles had used, but only for the purpose of slowing the ship down enough for the control surfaces to become useful. This enabled the Shrike to begin flying more as a regular airliner. It could bank and swerve enough to allow the pilot to bring it to the Uplift airfield under his or her personal control. The reentry point was selected to allow all this to happen without much fuss, or much fuel consumption. It was all also mostly automatic. In this case, Adam didn't have to exert himself very much at all. He and Ana mostly kept calling out "Nominal" when prompted to do so by the computer that was doing the flying. The Uplift field came into view at precisely the proper moment. Adam did assist the computer with touchdown and braking. Ana talked him through the taxiing, as well as engine shutdown. There was about a 30-minute rest period while the crew shut down the various systems, and the body of the Shrike just shed much of its exterior heat. All in all, Shrikes were, and remain, a bit of an aeronautic miracle. This was routine for Ana, but really a huge relief for the less experienced, or even unexperienced, crew. When they were cleared to disembark, more than a few tears were held back as they exited their craft.

Out on the tarmac, a relatively large contingent of Uplift employees were there to welcome them home. Adam seemed a bit annoyed, but good manners ultimately triumphed, and the crew was escorted to the interior of the main hanger. A little party had apparently been improvised, again with a head table and podium with a microphone. Already seated was Nathaniel Floatingfeather, there to welcome his interns, and one Frances Thorne, CEO and very proud father. Banners with "Welcome Home!" and "Congratulations!" were hung around the huge space. Standing behind the podium and yelling, "Come on in! Have a seat up here with me, you four!" was Frances, his face beaming. A lunch of fish and Shrimp Po'boys was served. Words of praise flowed freely. The little ceremony ended with presentations of Certificates of Qualification as Uplift pilots were presented to both

Adam Thorne and Ana Kozelka. Manjeet was certified as a qualified Flight Engineer. Mattie received nothing whatsoever. She didn't seem to mind, though, nor did she think she should. She did at least know she was now a confident and an appreciated companion of one of the richest and most motivated young fellows on the planet. She wished her parents were still alive to see it.

Nothing about the rest of the summer would approach this one crew's experience for any of that summer's crop of interns. It was all routine from that point on. Manjeet and Mattie wound up working closely with Nathaniel Floatingfeather dealing with Flight Crew equipment and design problems. Juni seemed totally enveloped in her computer design work. Adam, however, was only available to his friends in the evenings, and spent most of his time right at his father's side. You might find the two anywhere, from the hangers to the control tower, to any of the many halls and departments in the headquarters building. Weekends would usually find the various members of Number 9 at the Thorne "mansion," behaving like any group of college interns might, save for the lack of drinking and sex. Adam made it a point to fulfill his promise to Mattie, who got to drive a lot of truly top tier cars around that track behind the house. As a matter of common knowledge, no one who ever had an internship at Uplift Aerospace is known to have commented on their end of summer questionnaire that they thought it had been a bad experience. All observers that I know, however, felt that the experience of all of the "Nueve Mexicanos" that summer seemed even better than most. It's an inescapable conclusion that having Adam Thorne rooting for you might have led to some bias by the company toward his chosen companions. But not adding people he trusted to his professional circle would have complicated his path forward. As will be seen, it is hard to imagine he could have accomplished any of his lofty goals without the very interns he decided to choose this particular summer.

No one he ever decided on would ever abandon him, even though staying with him would be an extremely demanding task.

Chapter Six

Big Boy Pants

*From diary entries of Adam Thorne and Matilda Orona transmitted
to Uplift Aeronautics from Earth Orbit, Elsewhere, and various
interviews..*

The beginning of Adam's sophomore year at UNM was obviously greatly anticipated by everyone in the Neuve study group, especially after such a remarkable summer. Mattie moved all of her worldly possessions, sparse as they were, into Adam's small but luxurious little cottage. Juni moved her things into Manjeet's room, making it seem somehow less mundane. Mattie, however, continued to seem a bit aloof, more roommate than friend to the other three residents. This is not to say that she didn't voice her needs and concerns about anything and everything to her roommates whenever she felt it necessary. She just didn't consider her roommates as anything but roommates. Neither did she refrain from riding to school every day right alongside Manjeet and Adam. Juni continued to ride the bus though, which was a bit perplexing. She always said she just needed the time alone to gather her thoughts. The coursework was rather more advanced for everyone, as one would expect. Number 9 resumed in the same room at the library, and at the same early hour. Everyone was alert when classes started to scout for potential nuevo neuves, whether freshman or sophomores who might have improved since last year. Adam continued close e-mail correspondence with his father, in case there might be any new Uplift related ideas they each might have come up with before they could be together again.

In addition to regular coursework, all four housemates also took on some on-line classes that were offered by Uplift. These involved several pilot and crew certifications that they hadn't already earned.

Adam stood pat where he was, but Manjeet, Juni, and Mattie thought it wouldn't be a bad idea to do the coursework required to work towards pilot certification for the Shrike spacecraft. There was no time or opportunity to work on the simulator portion of their licenses, but they were guaranteed to get simulator time next summer when slots were available. There was a huge amount of content to complete before then, encompassing navigation, communication, protocols, and the like, and it took up a good deal of whatever spare time they might have had for their entire sophomore year. Most of this work took place online on the desktops they kept at the country club place where they were living. This year, they started to refer to the place as "Cockpit Cottage" or some variation thereof. The place was still strictly off-limits to anyone else and would remain so for the rest of their time in college. Nothing else of any great import occurred that year, which is not to say that every year of any student's education is insignificant. There were another few students that were invited to join the Nueve Mexicanos, and all of them were accepted as interns at Uplift the following summer, in 2081. As expected, Adam's roommates all became licensed and certified as Shrike pilots that summer, and each made their first flights working towards possible employment in that capacity when they were done with school. These 2081 summer flights were fairly routine Uplift missions, and to Mattie's great relief, none of them involved fish. What everyone unfailingly remembers about that particular summer is that Adam and Mattie seemed to be spending more time together and were definitely on friendlier terms. Those who were especially observant also noticed that they consistently drove home together and that Mattie was not rooming at the apartment complex. Apparently, Frank Thorne was getting onboard with Adam and Mattie as a couple. Juni and Manjeet were also rumored to be seen holding hands on a consistent basis. No other public displays of affection were to be encouraged, at least not where other Thorne college students could observe them. Incoming freshmen all were quickly made aware of the existence of the group, but they were always told by the upperclassmen to not bother soliciting anyone for admission. "If they want you, they'll tell you," they were told. "Otherwise, just do the best you can on your own." Oddly, Adam seems to recall that no one outside the group ever really figured out his true identity. Neither did anyone try to kidnap him, blackmail him, or rat him out to the press. After the usual four years as undergraduates, by May of 2083, the time had come for the original Nueves to graduate with their BS degrees in Aerospace (and Computer) Engineering. Manjeet was the class valedictorian in the former and Juni in the latter. Mattie was number two behind Manjeet, and Adam number three. At the graduation ceremony, Adam was called up as "Adam Thorne," which actually solicited a few gasps. This was hilarious to the now majority of graduates and spectators who had known for years that Thornhill was just a pale alias. The following day, there was a real celebration at Cockpit cottage. As far as anyone could tell, that sort of loud partying had never happened before in the stodgy upscale neighborhood they lived in. The story goes that the neighbors were initially not too pleased about it, at least until they were allowed to see the Maserati exposed when the garage door was opened. The event finished up as a full-blown block party. No one asked Adam his name.

In the weeks that followed, the residents of the Cockpit began to solidify plans for what was to come next, in their lives and their educations. Mattie, Manjeet, and Juni all expected to continue on with their master's degrees. Adam, however, shocked everyone when he announced that the only reason he had even bothered to get his bachelor's was that it was the minimum requirement for employment at Uplift Aerospace. He had already been offered a position as Executive Assistant to his father, and he had accepted. He also reintroduced the idea, which he had told them all about before, that they should forgo their own advanced degrees and follow him to Uplift as well. They could then go on to get higher education online. To his utter astonishment, all of them declined. Adam was fortunately not controlling or demanding enough to be offended by their decisions, far from it. He wished them well and offered them the use of his little mini mansion for as long as they needed it. Whether he questioned his own judgement is not known, but in the end, he found he had no cause for regret. His near-sibling roommates were also invited to visit him and his father whenever they deemed it necessary. He would keep the Maserati, though. He also made sure they had full scholarships for as long as they needed them. Mattie was initially somewhat ambivalent about accepting this charitable offer but wound up accepting anyway. Mangeet and Juni decided that they had saved enough money out of the allowance their parents were sending to go in for an extremely used Subaru for the three of the remaining Cockpit grad students to get around. Number 9 would be left to the underclassmen as a going concern. Mattie recounted that after these things were decided, Adam invited her to dinner the night before he returned to T or C. She accepted, and they took the Sandia Tramway at the base of the mountain which defines Albuquerque to the wonderful restaurant at Sandia Peak. Over lobsters, he confided that if she still wondered if he had the hots for her, his research had proven her correct. Just so she knew, he said.

Mattie confided that she had long had her suspicions that his internal thermodynamics were quite active. She apologized had not made inquiries about this particular subject, except for that one time on that maiden flight together. They were both so busy, she said. Besides, why would an obscenely rich kid with the whole world soon to be at his feet ever choose an upstart orphan with nothing to offer?

"Why would I ever choose anyone else?" she said he asked her. "If the world in front of me didn't have you in it, I'd have to go find one that did. My research indicates that only time I'm happy in this world is when you're around me. You must have come to the same conclusion, I think. You always seem to know what I'm thinking, and yet you never blush. That's fascinating to me."

"I always know what you're thinking because you do blush," Mattie replied. Adam probably blushed when she said this. "Look, I know we can't really be together right now, I'm still in school. The time will pass though, have no fear. But I'll come down whenever I can. Manjeet and Juni will bring me."

"Mi casa, su casa" Adam said. "You know, I have some pretty big plans for my life. I have to get started, and right now, I'm afraid I can't imagine doing any of them if you're not with me. But I know I may be asking too much."

"I'm afraid I have some plans of my own as well," she said. "I think my plans will be useful to your plans too if you can be patient and wait for me. When the time comes, I'll be interested to see what I can do for you. As long as you let me drive your car once in a while, that is."

"I love a woman who gets bored flying spacecraft," Adam whispered loud enough for Mattie to hear. He drove home to Truth or Consequences the next morning.

It was to be two full years before Manjeet, Juni, and Mattie were to complete their educational pursuits. They all came south to Uplift in the intervening summer breaks, and multiple other holidays and weekends. All of them assumed flight duties on those occasions and earned some rather good paychecks in the process. Adam and his father were closer than ever, Adam knowing full well that he was purposefully being groomed to take his father's place one day. Frances knew it too, and he made it very clear to anyone who would listen that his son was in training for the job. He even specified in his will that all his assets, as well as the position of Chief Executive Officer, would go to Adam in the event of his death. The surprising thing is that no one appeared to contest this. It was a case of family privilege, and there may easily have been candidates outside the company who would probably be better qualified. But this was a Thorne family business, after all. Besides, Adam was well liked, and evidently well respected, despite the fact that he was only 22 when he returned from school. This was one of the advantages of being an heir, and a very wealthy heir he would be. This transition was expected to occur sometime in the very distant future. Of course, as everyone now knows, it was not to be very far in the future at all.

On February 12th, 2084, Frances Thorne was in one of the propellant storage facilities near the main hanger off the Uplift flight line. He was inspecting a new pump system for loading propellant into the shipboard tanks that powered the small reaction control jets on every Shrike. The propellant was called Nitrogen Tetroxide and had been used for this exact purpose since early NASA days. There had already been one famous cabin leak on the Apollo Soyuz mission back in 1975. This occurred in the Apollo cabin right before, or perhaps right at splashdown. It briefly incapacitated Pilot Vance Brand and eventually resulted in lung problems for him, as well as astronauts Tom Stafford and Deke Slayton. Ultimately, Deke Slayton, a living legend and one of the original seven Mercury astronauts, eventually developed terminal brain cancer as a result of this exposure, which ended his life in 1993. This was due to only a very brief and limited exposure to the gas. Not so for Frances. The other staff had stepped outside the storage shed to check some gauges on the outside wall. Suddenly, one of the propellant feed lines broke, sending the chemical spewing into the closeted air at a significant rate. Alarms sounded, and the automatic door slammed shut, leaving Frances trapped inside alone and with no emergency breathing equipment at his disposal. The crew was able to shut down the flow of the gas from outside, but this took about two minutes. It took emergency air pumps inside another five minutes to clear the air enough to allow access back inside. But his employees knew that Francis had no chance of surviving such a leak. They were hysterical and devastated, and they were right to be so. Frances Thorne breathed no more. Oddly, it was the first employee fatality in Uplift Operational history. There had heretofore been only minor injuries at Uplift and none during actual flights. But this first death was the ultimate death and was crushing to everyone.

Adam was in his father's office at the time, reviewing projected mission requests, he said. The blaring alarm from near the flight line was one he had never heard before, nor had anyone else. He always had a fair idea of what his father was doing at any given time. He seemed to recall his father would be doing some sort of inspection about this time, but it never occurred to him that he would be so hands on about it. In retrospect, days later, he realized that, well, of course he would. He threw down his papers when he heard the alarm and sprinted towards its source. He arrived with all the other designated first responders. He asked what had happened, and the pale work crew told him what they thought had occurred. "Has anyone informed my dad?" he asked. The agonized men took a few tearful moments before they could mouth out, "It *is* your dad!" Adam simply collapsed, having instantly fainted, then came to on the ground sobbing. So was everyone else at the site. He barely recalled what the rest of the day was like. But Frances' body was recovered, the morgue was contacted, and the media was informed. Frieda Jojola, Frances' secretary, took Adam back to the mansion that was now to be his home. His alone. She comforted him as best she could and reminded him that the company was now his alone as well. She told him that he would be expected to pull himself together as soon as he possibly could. He was head of the household now, and it was one of the biggest damned households in the world, the Uplift household. As his first painful task, he made two phone calls. The first was to the Uplift press liaison on duty that day. The second was to Mattie. She said the Cockpit Cottage crew would be right down. Manjeet had a Subaru.

The funeral was one week later. Frances would be buried in Las Cruces, his hometown. Adam was faced with intense media scrutiny for weeks after his father's death. Financial markets wavered, as investors expected Frances' stock portfolio to go on the market. The idea of Uplift being in the hands of a mere child with a bachelor's degree was seen as entirely ludicrous by Wall Street. But Adam went on the news and gave a press conference which reaffirmed that he was well capable of taking the reins and had the full confidence of upper-level management and the entire staff. None of the staff they interviewed for the record disagreed. It didn't hurt that it was actually true. Mattie and his two other former roomies stayed at the mansion for four days after the funeral. Adam was still consumed with sorrow, but it was strictly due to the fact that he had lost his father, and not because he was fearful about his future. He actually seemed a bit upbeat about that part of things. In a way, it was as if the need to satisfy his father had previously suppressed his inner ebullience. He was always afraid of his father finding some flaw or weakness in his thinking or his actions. Frank could certainly be said to be guilty of making Adam feel that way. He also always had that engineer's need for control, of Adam and of himself. Maybe he had passed that same need on to Adam as well. Adam sometimes did feel that maybe he was part robot somehow, too much insensitive to others. But with Frances gone, it was perhaps time to try and be a human being, a complete man. He was with the three people on the planet he valued most that weekend, and he decided to make it count. He cooked monumental dinners for the four of them, all by himself. He wanted to know what plans they had for their lives, and if they would include him in those plans. He wanted to know when Juni and Manjeet would be getting married, and if he could help with the wedding. They all bared their souls to each other the whole time they were together. That first night, he asked Mattie to stay up with him for a while.

Of course, the exact conversation was not recorded, but those of us who worked with him were able to gather that Adam begged Mattie to come stay with him and finish her schoolwork at Uplift. She apparently again begged off, stating it was his fault for lighting the flame that drove her to learn as much as she could as an aerospace engineer. It was Adam's father's fault, too, for creating one of the great colleges on Earth to teach her. But then again, it would only be another year or so. If he could wait, then she might let him marry her. "Deal!" Adam was said to have blurted out. "Wait, what?"

"You heard me!" she said. "Sounds like you didn't even have to think about it."

"Apparently my heart has somehow become verbal", Adam said, in some sort of enraptured whisper.

"We'll have to give it something else to talk about, I think. Do you get to sleep in your dad's bedroom now?"

"Yeah, I suppose I do. Let's see if I can remember where it is." They found it together.

The following Monday, Adam had composed himself enough to convene his first Board of Directors meeting. Though he generally sat near his father's elbow on the right side of the long table, this was his first time at the head. Frieda Jojola, who was now Adam's executive secretary, performed the visual presentations as needed, and also recorded minutes. Being so competent and experienced, Adam knew this was a job she could do in her sleep. She had gone over the agenda with him the weekend before. He wasn't nervous or shy about running meetings himself. After attending these gatherings once monthly for the past year anyway, he was entirely familiar with the proceedings. It all started with a summary of the last month's budget; income received from various sources and expected outlays for the coming month. A review of contracts with the US government, especially NASA, occupied the next thirty minutes. Uplift Aeronautics had turned its usual 500 million or so overall monthly profit. The 126 missions flown this month were an entirely typical number by this time. Adam noted that with a total operational fleet of 150 Shrikes in the home fleet, this left 24 that had not even flown that month. He was reminded that a further 20 birds were on loan to other countries. He then announced that he viewed this as a suboptimal use of the company's resources and pledged to reevaluate this problem. On a more somber note, he called for a summary of the investigation into his father's accident. All that had been surmised thus far is that the work crew had failed to do an inspection of the pressure lines prior to the testing phase. It could not yet be determined whether or not the lines had been pretested at the factory or if they had somehow loosened while or after they were installed. The investigation was ongoing, but Adam directed that a more stringent safety checklist be developed for all equipment not manufactured by Uplift directly. Company produced equipment already had ironclad guidelines, as Frances had always demanded.

The meeting now turned to the topic everyone had nervously been expecting. This was to take the entire five hours remaining after breaking for lunch. This content is included in our narrative because of how it gives us all at least a small peek into what was going on in the mind of Adam Thorne at this pivotal point in his life. What, the question was, would Adam wish to do with the company now that he held the reins? Would it be business as usual, a whole new direction, or some combination of both? He had been privy to meetings with his father's accountants over the last year and knew the company had a stock of financial resources far exceeding its ongoing debts. His father's, and now his, personal wealth exceeded approximately 280 billion dollars. This did not count his property holdings, almost all of which were in a 50-mile radius of where they were now sitting. Would he stand pat, or diversify his interests? They were sweating with anticipation as he started speaking. Adam stood up in order to use the screen and the obligatory poker provided. The conversation was recorded, an old company tradition that Adam would continue.

"As we all know, my father's contribution to this country and the world have been transformational. Never has the human race had greater access to the orbital environment of this planet as it has today. I can't help but feel it is as great a contribution to the world as those accomplished by the likes of the Wright Brothers, or Eli Whitney and his cotton gin, or even Fulton with his steam engine. We have gone even farther with the development of our Shrikes, as well as the technical support we provide for them. But now, these blessings have been passed to the next generation of their stewards. Most of you represent somewhat the gradual transition from the last generation to a newer one. I implore you to help me make that transition. I must confess, I have a substantial number of ideas about what that transition should look like. This seems to me to be a time when access to space will resemble the transition from when those feeble first railroads exploded into universal access to transcontinental travel. We will be the Cornelius Vanderbilts of the late 21st century? We certainly could be if we use all of our vast resources and boundless initiative!" This all came as a bit surprising to those present in the room. They all knew Adam was bright, but he was just so damned young, and had always been rather quiet all these years at his father's side. Who knew he could motivate like this? Then he continued. "I must say, my father has run a very tight ship! And a happy ship too. Quite a remarkable achievement, given the history of American businesses over the centuries. Fortunately, after our recent wars and the general reacquaintance with the principles of democratic rules and income equality, I will gladly continue to ensure all Uplift employees are paid at the highest levels anywhere, and all necessary needs for our health, safety, and education here at Uplift will continue to be met. I will not shy away from paying lawfully levied taxation by the US and New Mexico state governments and will strive to protect the Earth's environment to the utmost possible levels. In other words, I will emulate my father in every way I can."

The board members applauded, and at least seemed comfortable with this first speech, although it did seem a bit portentous and even somewhat cliché. He continued. "In point of fact, I intend to begin using my personal wealth substantially more freely than my father in the pursuit of humanitarian causes" The financially minded members present did appear to become a bit squeamish at this announcement. Adam appeared to expect it. "Don't worry, I'm not going to go over the deep end and become some weird hermit like Howard Hughes or anything. To start, I'd like to address helping the homeless population, especially here in New Mexico. At least we're not as bad as a lot of places, mainly because most people think of us as generally destitute in the first place. It's also thought of as hot and dry and generally inhospitable to human life." There was general chuckling. "We New Mexicans don't go too far out of our way to dispel this nonsense though, mainly because we really don't mind our privacy. We content ourselves to let most relocating Americans wander their way to Arizona, even though they are even hotter and drier than we are!" More laughter. "We do have our fair share of homelessness of course, not to mention poverty and crime. "So, what I'd like to do is begin providing inexpensive housing, food availability, and job training for our disadvantaged people here in the state. I know these are words we hear all the time, from every sector of the country. I'd like to put some real money into it, though. I know a number of developers have designed inexpensive modular houses which can be constructed quickly and easily. I'm not talking trailers or tiny homes either, but houses that can accommodate large families, with a lot of storage, even garages. They should be on true lots with backyards, have adequate kitchens, bathrooms, and laundry. They should be in neighborhoods with paved roads, stores, electricity, clean water, all of that. And we should be willing to entirely fund whole communities in this way." Not much of a response of any kind was recorded with this bit of speechifying.

"Governments will hopefully get on board with this eventually too, but the majority of the funding will be from me. We'll need the communities to have fire and ambulance services, as well as police. That will be the government's part. But I want to ensure that in exchange for living there, every household will be offered an opportunity for someone to be employed, with training provided right there in the community we build. I feel sure that there will be a good many people who will come and work with our company eventually. But I'm not talking about a company town situation. Shoot, we already have one of those right here. You can stay in this new little community, or communities, if you keep your place up, try to work if you can, or at least get government support, and obey the laws. Otherwise, you're out. Now I know this is idealism taken to a nearly preposterous level. All these people are likely to need health care, counseling, psychiatric care, the whole deal. We will probably not reach a lot of the goals I'm setting; I know. But I can't stand seeing such a large portion of humanity, including a lot of Americans, left outside to waste their lives away. Our politicians try their best, but the richest of them really don't care who lives or dies, so long as they stay rich. And that's not how I intend to live my life. I assume I'll stay rich, but I can at least spread it around a bit. Now, I thought about naming this initiative "The Thorne Foundation," but that seems a bit too prickly. I think I'd like to go with "The Uplift Foundation." That's a bit corny too, but we have to call it something. So, Frieda, if you can brief the legal department guys to develop some proposals, I'd like to hear back within the next couple of months. Alright everyone, that's all I've got for now. We'll meet again one month from today. I'm afraid I have a lot more I want to explore, but those things are definitely more in our usual line. Otherwise, let's keep performing at our usual remarkably high level. I do hope you'll keep me around. I can't believe how lucky I am that you're all still here." The meeting was dismissed, leaving the attendees as much perplexed as they were inspired.

Even after a week, the Board was frankly ambivalent about what Adam had imparted to them that day. It was good to know that at least he was familiar with the basic workings of the large corporation that he now controlled. On the other hand, he used a good bit of their company time to discuss what were basically his personal agendas. Opinions were mixed regarding his proclaiming that he intended to become a philanthropist. But since it would be done with all his own money, why take the time to discuss it with the board? No one could tell, but it seemed to definitely represent a true change in the focus of what had heretofore been, basically, a really large freight company. The various department chairs, in the days afterward, began whispering among themselves that Uplift appeared to have a new eccentric billionaire on their hands, akin to Jeff Bezos or Elon Musk. Father Frances had been a laser-focused straight shooter, always with Uplift's mission on his mind. They tended to hope that this focus would never change. But further reflection, especially among the business-minded types, reminded them that, generally speaking, businesses that never evolved or refined themselves eventually went extinct. Business principles were not all that far removed from Darwinian ones. Thankfully, no one chose to depart the place after realizing what Adam was so gently implying. Uplift would, in fact, begin to realign its priorities. The big question would be how. It wasn't long before the directors' suspicions were shown to be correct. In the weeks that followed, Adam had learned that a large American hotel chain, under a subsidiary name of "Endless Horizons," had bought out a patent for an orbiting space resort. A number of these types of stations had come and gone over the past fifty years or so. All of these ventures would fail rather quickly, only to be replaced by another company's version. This newest company could only be described as a" knock off." There were enough little changes and innovations to entitle the new designers to their own little patent, but the same role would be played by these newcomers as what had been attempted before. This new hotel would obviously be available only to the super rich, and rather brave ones at that. The designers had planned to secure funding for the project via all manner of questionable means. These included such schemes as lotteries, stock sales, and so on. But none of these would come close to securing the billions of dollars that would have been necessary to get the thing built. The main problem had been the cost of all the rocketry and tech needed to put the building materials into orbit. The advent of Uplift, however, with its large fleet of Shrikes, offered a ray of hope to the project developers. Even though Uplift had been in operation for all these years, it still would take time for them to get the necessary resources together. Other big orbital space firms, such as Space X, Virgin, and Blue Origin, had been too busy with other priorities to commit to such a project. NASA had grown tired of the concept by now as well and also declined to participate. Frances Thorne had apparently been mildly interested earlier but never fully pursued it.

The concept was ambitious, especially at the outset. The hotels were a primordial development of an idea first conceived by none other than Werner von Braun, the former Nazi V-2 designer who went to America after World War Two. Here, he helped the Americans to begin designing the missiles with which the Cold War would be perpetuated. He became increasingly competent as the world's greatest rocket designer until culminating his genius with the Saturn V moon rocket just before his death. He had an oddball idea along the way to suggest the so-called wheel in space concept. This concept was fictionally popularized in such Movies as "2001, A Space Odyssey" and many others. In such a structure, a large tubular shaped craft, with a hub in the center and radiating spokes, would revolve at a moderate speed, allowing for artificial "gravity" to be created on the inside of the outer rim of the wheel that would assist the occupants to live in increased comfort. Long experimentation since that time had proven that while helping humans to lightly get pushed towards the outer internal edge of the wheel in some fashion, people would still become dizzy, nauseous, and not really able to function. Still, for this hotel design, it was felt to be a large enough wheel that, with a slow rotation rate, the vertigo would be tolerable. The simulated gravity it generated would approximate that of the moon. Variations on this concept had already been flown, using a variety of complex methods to build them in orbit. The proposed new one wasn't particularly different.

"Endless Horizons" had designed a central hub for the ring that was circular in shape. Four cylindrical struts would radiate from the hub that were cylindrical in shape. These would contain elevators to take people back and forth to the outer tubular ring structure that contained the rooms and common areas for guests. The central ring has a slot-shaped landing bay where guests could be flown up to and down from the hotel. The shuttle vehicle could dock to an air lock to embark and disembark. One of the problems the program originally had, which gummed up the whole plan, was the lack of any development of a shuttle craft for ferrying passengers. Otherwise, complete plans had been made for all aspects of the structure. Special space compatible robots were already designed and built that could undertake the Extra Vehicular Activities (EVAs), or spacewalks, that were needed for the build. Actually, large conventional rockets would still be needed to construct the central hub. Again, this would accommodate the docking facility. But office space, oxygen and water supplies, heating elements, and fuel for the rockets needed to impart the spin and attitude of the orbiting palace would be located here. For the other structures, though, such as the framework structures and the rooms themselves, Shrikes could be used extensively to get everything into orbit and assembled much more cheaply than any means used before. Expanding frame structures had been devised and utilized since ISS days and were not difficult or expensive to build. They would be engineered to fit into any Shrike's cargo bay. The habitable areas of the outer ring were even more simple. For several decades, designers had envisioned, and often built, what amounted to inflatable playground-type equipment that could take nearly any shape, be airtight, resealable, and micrometeor protected. These could be tightly folded in compact shapes, taken to the proper place in orbit, connected, then inflated. The plan was for thirty of these structures to be taken aloft and assembled into the necessary ring structure. Once filled with breathable air, the interiors could be fitted out in any luxurious way specified in the design. There would also be four air lock structures, which would have escape vehicles docked along the ring for emergency evacuation. Shrikes would be purchased by the company exclusively to be used for that purpose. One side of every compartment of the ring would have a large solar panel attached to provide power to the hotel.

Adam liked the proposal when it was presented to him. He observed that the cylinder compartments were divided so that one-third of each tubular section was devoted to a hallway that could be used to move room to room by the guests to get to their assigned suites. The suites themselves would occupy the remaining two-thirds. This would make for some oddly shaped compartments, as the sections would be tubular in shape and only about 30 feet in diameter. He suggested the shape either be changed to cargo-type rectangular sections or that a second course of tubes be positioned parallel to the first so one could be used as a hallway and place to run wiring, ducts, and so forth. The other would house the actual suites and compartments. This would not be particularly more expensive, as there was adequate cargo space aboard every Shrike to accommodate the new components. The designers said they would consider this. Reviewing the whole proposal, he projected the whole contract could be accomplished at a cost of 8 billion dollars, providing that all materials and management of the project was provided by the owners. Shrikes would provide transport to orbit of all necessary construction personnel and materials, as well as flight crews and use of the robotic arms present on every Shrike. Other associated costs could be negotiated as the project progressed. As this was tens of billions of dollars less than what had been projected when the project was first conceived, the offer would be formally drawn up. Adam agreed to propose it to the Directors when all was prepared.

Adam also reviewed all of Uplift's current and ongoing projects. Odd as it seemed at the time, the US government continued to fund the clean-up of the orbital debris that had accumulated for much of the past 120 years or so. Most of this had been launched by the United

States and ranged from spent rocket engines to dead satellites to shrouds and plain old shards of metal. It was a miracle that no outright crashes with active spacecraft had ever occurred, either manned or unmanned. Frances Thorne had contracted out for this activity about five years before his death. At any given time, at least one Uplift Shrike had been in orbit collecting this copious litter in space. The rear bay robotic arm was most often used to scoop the trash into the cargo bay for return to earthbound recycle bins. An extendable net was also devised for the smaller floating objects, which would snag them and haul them in. Sometimes, specialists were taken aloft who could reactivate or repair the younger satellites. A few of the larger objects, which could not fit easily in the bay, had the handy small solid rocket engines attached. These were fired off either toward an eventual plummet into the sun or towards an uninhabited portion of an ocean below where it could burn up harmlessly in the atmosphere.

Next on the list of services Uplift provided, as has been mentioned, was the launching of large blocks of plastic which had been retrieved from the oceans of the world. After the political shakeups of the 2020s, the nations of the worlds had become united in reducing the production of plastics in favor of biodegradable alternatives. Most seafaring nations were now involved in dragging the oceans and seas in order to collect the hundreds of thousands of tons of plastic waste that had been dumped there over the past century and a half. These were often recycled, but if this was impossible, Uplift was also given a contract to shoot the stuff off the planet. In the past 20 years, the removal of plastic from the environment at large had begun to outpace the amount that was tossed into it. There was some controversy about whether or not burdening the sun with plastic was a good idea, but it was generally agreed eventually that there wouldn't be much of an impact on the star's functioning. Another initiative Adam began to explore was the firing of the world's nuclear waste into space as well.

If packaged properly, with no leakage of the stuff into the cargo bay or anywhere else while loading it up, it seemed to be a better idea than burying it underground in various sites around the world. Of course, people had always been nervous about the idea of radioactive materials orbiting over the surface of the planet, and this wasn't ever likely to fade away. The idea was to shoot the spent Uranium, Plutonium, and such into the void as soon as orbit was reached. Maybe it could also be fired at the Sun. This would need further study in regard to safety as well. Hmm. Another idea would be to make the planet Mercury into a huge radioactive waste dump. No one was ever likely to want to visit that hell hole of a planet anyway. But perhaps humans should go there first to make sure there really wasn't any life there which would be impacted? Likewise for any other planet or moon. Or perhaps the stuff could be shot out into the infinity beyond the solar system. Not very neighborly, but who knows if any neighbors were out there. This was all food for thought in Adam's brain during this time.

A question that has always pestered the conspiracy theorists out there is whether or not Uplift had ever been involved in military type activities. Obviously, this would be classified information, and no one in the know has ever been granted permission to discuss or disclose any facts related to these questions. It is still a sad truth that the free countries of the world continue to have enemies in the not-so-free ones. One can imagine how the development of easy orbital spaceflight have annoyed these potential, and also some very real, enemies. Spy satellites remain proliferative from numerous countries. Who can say that potential anti-satellite weapons, or even anti-nuclear missile weapons, weren't being launched clandestinely from New Mexico? The answer is that no one can. The question has in fact wound up being a driving incentive for a modest amount of denuclearization by the world's so-called superpowers. It has made nukes seem a little bit more of a potential waste of money. Also, the lasting realization that no war is worth destroying the Earth over has reduced, but not eliminated, the production of nuclear weapons. Has Adam Thorne ever had to ponder these questions? How could he not? Who knows if he ever came up with any answers.

One other going concern that had engaged Uplift's services for several years was the orbiting medical research facility that had been assembled in space using a fleet of Shrikes. This laboratory had changed ownership twice in the last three years and was currently known as Queen of Heaven Orbital Research Laboratory. Obviously owned and operated by the Catholic Church, Uplift continued to provide needed services for personnel transport and support, resupply, and facility maintenance. Adam kept close tabs on its activities and discoveries and found himself wanting to take a more active financial role in its operation. He began to realize that the more humans increased their presence off of planet Earth, the more medical services needed to follow close behind. Space X had found this to be entirely the case as they increased their attempts to inhabit the planet Mars.

Nathaniel Floatingfeather, Uplift's longstanding Chief of Operations found himself becoming increasingly enthused with Adam's energetic approach to his new responsibilities. He hadn't realized as yet that the company was becoming rather predictable in its approach to business and spaceflight in general. This came as a kind of cold shower to him when he did, and he opened his eyes accordingly. He began to sense that Adam was quietly contemplating a tectonic shift in Uplift's business plan and was putting out feelers to decide when to apply the gas. Nathaniel eventually consented to help in whatever way he could. In his spare time during those first three months, not that there was much time to spare, he helped Adam write what turned out to be a master's thesis that he submitted to the UNM

Aerospace college. The topic was just a summary of the design and manufacture of the large aquarium tank which he flew to space on that solo Shrike mission on the summer after his freshman year. Though his father had provided the actual materials and facilities to make the tank, Adam was the primary designer. Nathaniel used his influence to have Adam enrolled in the master's program by proxy, and the degree was awarded forthwith. Adam was rather less than thrilled about this, feeling Nathaniel had applied undue influence on his behalf. He called it an "Honorary Degree" and insisted that no further personal academic advancement was desired, or in fact required for him to advance any further academically. Nathaniel felt that a master's might give him more credibility in holding his position, but Adam knew that this simply wasn't needed. American business could be a simply hereditary proposition, and very often was.

By the end of the year 2084 some of Adam's plans began to bear fruit, or a few blossoms anyway. The space hotel proposal was accepted by the company, as well as the FAA. Uplift was awarded the construction support contract, with February 1st as the target date for the first of dozens of Shrike missions to commence getting the thing built. Furthermore, at this time Mattie, Manjeet, and Juni had all completed their doctorates and moved back to onsite housing at the Uplift America Station. Mattie moved in with Adam, which surprised no one. Manjeet got his own small home on the Uplift grounds. Juni elected to move back with her parents. She informed them that she was converting to the Sikh religion at this time. Her mother and father, both being very open-minded scientists, made no objection, happily feeling the decision was hers alone to begin with. This being the case, the couple were now formally engaged, and planned to marry in May of that year. Adam, always one to follow Manjeet's example in all things, decided that becoming Mattie's husband was probably the best thing for him to do as well. He popped the proverbial question the week after Manjeet did. Mattie, however, urged a bit of patience. Though we have no transcripts of these conversations (obviously), we do have transmitted diary entries from both Adam and Mattie which can be cobbled together now.

"After I had moved in with Adam, I realized how comfortable I actually felt with him," Mattie wrote. "Even so, our interrelations had always been as engineers, using technical jargon and the like even when discussing nontechnical things. He always seemed to me more of a presumed superior, someone I had to please, or else he'd toss me away like a bad blueprint or something. We'd never properly verbalized anything like love or need for each other. I knew this was as much my fault as his. I didn't want to seem like I was trying to suck up to him just to get his money or anything. I felt that if he didn't think of me as an equal, then there was no point being with him anyway. It was a long process before I knew I was in the right place being with him. Apparently, this was the best approach, but nothing is perfect when it comes to human beings." At last, Mattie had finally accepted in principle that they should be married. First, though, she asked him to hire her as a design engineer. She reasserted that she wanted to work with him closely for at least six months before accepting his proposal. He of course agreed, and graciously promised to begin speaking to her as an actual fellow human being and not as a "piece of technology." He also told her then that Manjeet and Juni had already been accepted for management positions in system integration and computer design, respectively. He was going to assemble a task force, he said, for a new project he was thinking of, and he now felt he had the right personnel available to proceed with it. The first meeting would be Wednesday of next week. Mattie smiled, and then kissed Adam goodnight in an entirely non-mechanical way. It was a satisfying conclusion to the nearly two years Adam had spent gearing up for what would end up being a very remarkable career.

Chapter 7
A New Direction, But Still UP

Adam's propensity for recording all of his staff meetings is especially helpful when it came to the very first meeting of his "Future Projects Task Force." It was, in many ways, a renewal of the old college "Numero Nueve" study group from the undergraduate days. Adam had used his prerogative as owner and CEO of Uplift and hired everyone now sitting in the boardroom on his own, with very little in the way of outside advice or consent. Not that no one there was qualified, everyone was, and extremely so. Plenty of other applicants for the myriad of the usual open positions were hired through the prescribed channels. For this group though, he needed people who he felt he knew and trusted completely. The members, now seating themselves at the long conference table, could sense that something was a bit odd when they arrived. They had all become rather used to this strange air of mystery from those college days, and every single one of them must have been a bit amused when they found out who else had been appointed to the team.

The former occupants of the "Cockpit" were all there, of course. Juni, Manjeet, and Mattie would have been aghast if they hadn't been selected. Others had been study companions as well but had not been mentioned in this narrative before simply because it would have clouded an already turbulent story. There was old Nueve, but new PhD Azaan Nazari, a Tanzanian man who had specialized in Spacecraft propulsion systems. He was 25 when this meeting convened. He was six foot two and weighed in at a solid 230 pounds. He was perhaps the most affable man in the room, at least until called on to speak about his field of interest. Then he was fiery enough to perhaps orbit a spacecraft with his fiery intensity alone. He had originally been hired to refine the Shrike's already useful Aerospike engines going forward. He also seemed to be very interested in alternate and theoretical engine types. Not a Neuve but a nuevo to the group, Awamila Hayat was Pakistani. She had been recommended to this task force by Nathaniel, who had interviewed her the year before for an advisory position. She was an astrophysicist and astronomer who earned her doctorate in the early development of the universe, all by the age of 24. She was 26 now. Her fairly new and unique discipline arose following the deployment of the Webb Deep Space Telescope in the early 2020s. Adam evidently felt this was a skill set he needed to include to drive toward what he was hoping this task force would eventually address. She was 5 foot four inches tall and weighed 116 pounds. She insisted on wearing the attire of traditional Pakistani females when at work, and was always calm and composed. She had a very commanding presence.

Next came Gregor Levinski. He was known to everyone as a specialist in systems integration and design. Also a Nueve, he hailed from St. Petersburg originally. His skill set could initially seem to be a repetition of Manjeet's, but it wasn't really. Gregor had always been more interested in life support type systems, including temperature and atmospheric regulatory equipment, electrical support, and water management systems as well. This also included less romantic systems like sanitation and waste management as well. He came to UNM from a different background than the Uplift set. He maintained a mild Russian accent and was not particularly patriotic when it came to his birth country.

Iza Markiewicz was a woman of Polish descent who had started out in Number 9 but veered off in her second year to specialize in robotics design and manufacture. She was 28 years old, 5 foot two, and 140 pounds. She had medium brown hair and green eyes. Her East

Philadelphia accent remained fairly pronounced. Mary Dupree was an Albuquerquean of color who was a specialist in materials and 3D printing-based manufacture. She was also a Neuve Mexicana but hailed from Detroit originally. She weighed 160 pounds with a height of 5 foot six. She had wavy dark brown hair (when she hadn't dyed it blue) and had a tough sort of back street demeanor. This toughness only made her even more endearing to the other members of the group. She also started out in Number Nine with the initial intention of being an aeronautical engineer. Her interests soon drifted towards materials and manufacture, but she continued studying with her original friends. This made the nine members of the task force that were all acquainted with each other, at least by sight. The tenth member was Dr. Peter Hosokawa, and Japanese immigrant to Hawaii who was a general and thoracic surgeon, having earned his degrees from UCLA. He was the oldest member of the group at 30 years of age. He had been the head of Zero-G research at the orbital medical research facility and was considered a definite prodigy when he started there at age 27. This was the "Queen of Heaven" station mentioned earlier, and he turned out not to be the best fit while there. He apparently had chosen to leave because he had found himself unable to conduct any truly pioneering work at the station. Instead, he was expected to continue only the types of practices that had already been developed there. Namely, this involved only the development of physical therapy techniques for the disabled and the treatment of severe pressure ulcers. Granted, this was useful work, but all other attempts at medical research were basically prohibited. Dr. Hosokawa was five foot eleven and 175 pounds. He was extremely assertive in all of his interactions with others. His parents had immigrated to New Mexico before he was born, but he had the appearance and commanding presence of a gentleman of Japan.

So, these ten members were assembled, and Adam's presentation began. "Welcome everyone!" Adam commenced. "Thank you all for making yourselves available to me for forming this task force! I really don't think I could have considered trying this if any single one of you had declined to participate. I couldn't have made any better choices for participants that might best help me meet the objectives I'm considering for this new project. I think that even researching what I have in mind is something that will require dedication, vision, and competence in fields that don't usually come together in this particular way. I hope I can pique your interest in staying here if you'll listen closely to my presentation today. Of course, if I fail, anybody here can decline to continue, and other less unusual duties will immediately be assigned to you. Your negotiated salaries will not be affected in any way, whichever way you choose. May I continue?"

"You'd damned well better!" Mattie grunted in a Mattie-like fashion.

"Excellent!" Adam countered. "I should say up front that Mattie's brand of frank interaction is precisely what will be required from each of you if you wish to serve on this task force. I am the nominal head of this group of minds, but I am most likely not the brains. This job will require honest and assertive participation from everyone here, every single person. If you generate any ideas or insight into any topic we discuss, you are obligated to bring it to our attention. You can do so in any way you see fit. As long as you don't denigrate or insult any other member, you are obligated to speak up. Does this meet everyone's approval?"

"Absolutely!" Manjeet shouted. This was unusual because Manjeet was traditionally quite taciturn in nature. Those who knew him well were quite amused, but knew he was indicating even he could become enthusiastic about Adam's ideas. After all, he was the most longstanding member of this formally informal group of old acquaintances. Usually, one had to try and deduce whether he agreed or disagreed with your suggestions mainly by observing his expressions. He was marking this change of perspective on human interaction with this forceful agreement. Everyone would remember it. "So, what have you to ask of us this fine New Mexico day?" Mangeet asked brusquely. The half-smiling group of people chuckled together, all looking attentively at Adam.

"Well, since you asked! Here goes." he began. "To start with, I would like all of you to be assigned at present to assist in the construction of a new orbital hotel we will be constructing this year." Moans were heard from everyone, and justifiably so. After such a grandiose build-up, this seemed extremely lame. "No wait, I know that's not particularly groundbreaking," he responded. "Now hear me out! I'd like everyone to pay particularly close attention to the methodologies and problems that arise with this project, and also to the solutions that are discovered. I think everyone who knows me has been able to deduce that I'm interested in space flight that is sustainable, with minimal need for resupply or assistance from the ground." This might not have sounded too impressive to an outsider, but it was a thunderclap to the group of knowledgeable colleagues in this conference room.

"Oh my God! Fish!" Mattie blurted out. How could anyone forget one's first spaceflight?

"Well yes, that was kind of an opening salvo," Adam admitted.

"For those of you who don't know," Mattie explained. "Our first summer as interns, Adam flew a huge aquarium of big fish into orbit. He just said he was curious if it could be done. Well, it could!" Many in the group looked puzzled. "He said it was about sustainability! Having a renewable food source right there in orbit! OK so wait, I'm not sure where this is headed. You want this hotel to have an endless supply of fish?" The rest of the room's occupants seemed ever more worried.

"Well, it may come to that, but no, that's not the point," Adam went on. "I am just really interested in finding out how long an orbital station can remain off this planet and be completely self-sufficient. I know the Mars colony is doing pretty well, but it still needs frequent servicing flights for personnel exchange and resupply. Frequent flights. I don't see that trend ending any time soon. I'm just interested in seeing if a self-contained spacecraft, which is what a station really is, can independently sustain a small crew for an extended period without any assistance whatsoever from outside. It's just in the interest of research, that's all."

Azaan Nazari, the deep voiced Tanzanian propulsion specialist, didn't hesitate in his willingness to speak his mind in full. "And why does an orbiting spacecraft need to have propulsion systems? The kind of craft you seem to be envisioning would need to be pretty large! But even so, a few conventional RCS motors would be all you need to keep it positioned."

"Well, as long as we're cutting new edges, who knows what we'll be needing!" Adam countered. "Besides, it's just research, so let's put everything to the test. If we don't investigate some of these issues, somebody else will later on. Why not us? Why not now?"

"Good question, Adam," Mattie said. "The problem is, you're not exactly known for being out in the open about your intentions, or what you really have in mind."

"Well, that's fair," Adam admitted. "In this case, though, I'll only have in mind what you all put there. I promise that anywhere our research leads will only be a product of our collective energy. I hope it's not a waste of your time, but that's how all human progress is ever made. Maybe it won't lead anywhere, but who can know that in advance? Research is generally its own reward from what I can tell."

Juni, who was always well known for her insight, again confirmed the willingness of the group to contribute. She, along with Mattie, was usually one to yell out the unspoken parts at the top of her voice, and usually without warning. "You want to leave the planet, don't you!? You want a big ship to just leave and stay away for as long as you can! I know you, Adam Thorne! That's what you're saying, isn't it?" Adam for once wasn't expecting what he should have expected.

"I can't imagine it can go that far, Juni," Adam finally said calmly. "But I want to honestly imagine how someone could one day. If we can't do it, maybe we can help someone else make that leap though. Really, many others already have gone off planet, but only to the next rock over at the most. Someone needs to take the next step, maybe even to the next star over. It has to be done with real people using real technology. Humans with real human brains have to figure out how. The very best real brains would be even better. And someone needs to do it before society makes the brains too afraid to try. So, what's the harm in pondering about it? But you're right Juni, that is what I'm thinking about."

Doctor Hosokawa put his two cents in. "You're thinking you need me to find a way to provide medical care aboard a spacecraft for a near indefinite period with no outside help? What a fascinating thought." You could hear his brain clicking. "We're not talking about an aspirin a day, are we? My God, do you realize what you're saying? That's everything from dental care to cancer therapy to any conceivable kind of surgery there is. I think you'll need way more than some decrepit old sawbones on a wooden ship from the sixteen hundreds. Have you really thought this out?"

"OK, let's not let all this get away from us!" Adam reiterated. "This is all just going to be simple concept management for now. But I must say I see I've chosen the personnel correctly. That's the sort of forward-looking realistic thinking that this task force is really about. You're anticipating the questions before anyone else is even asking them out loud. This is a "What If?" consortium. Juni? What if we could come up with an AI computer to take aboard with us? Gregor? What if we could convert water into Oxygen and Hydrogen fuel right on board? Awamila? What insights could we gain by observing the Universe from even deeper interstellar space than we have so far?"

"Interstellar Space?!" the astronomer cried. "You keep moving the goalposts here!"

"What scientist doesn't?" Adam said. "Look, there are all sorts of questions to be answered. More than we can think of at this little moment. We'll be needing more people. A nutritionist, for example. We'll need more than fish. We'll need vegetables and other protein sources. We'll need medicines, we'll need super long-distance communications and ways to invent and build things without help from the ground. What about waste management? What about gravity deprivation? What about radiation? You see what I'm saying? Somebody has to ask the questions. Who knows if we can really answer any of them? But I'm asking you to try. That's all. Why let somebody else do it? This is as good a group as any for these topics. For the time being, though, let's see if we can manage to help build this silly hotel. After that, who knows? What do you say? All those in favor?"

Adam was pleased to see that everyone had raised their hands. The expressions on their faces concerned him a bit more, he had to admit. Nevertheless, he slammed down his hand and yelled, "Sold." I'll be in touch with each of you by e-mail tomorrow with some suggestions", he told them.

Always true to his word, Adam e-mailed various small assignments to his new team the next day. The most pressing one was for Manjeet and Mattie, instructing them to meet him on the flight line in the morning for a company jet ride to Phoenix, Arizona. He had arranged to meet with the manufacturers of the stateroom units for the main body of the hotel project. These would be 3-D printed cylindrical units with connecting docks on either side. They would be assembled end to end with a small elbow kind of adapter between each "pod" to help curve the assemblies and form a large semi-donut-shaped structure. Each would have built-in windows to view the Earth below. The company was apparently taking Adam's advice to actually have two donut structures placed side by side, basically making one solid ring. The second donut would constitute a circular hallway with doors leading to the staterooms. There would be a separate doorway into each room from the hallway. In this way, you didn't have to pass through private rooms to circumnavigate around the structure. Also, the habitation pods didn't need to have space subtracted to make room for passageways, although some of the rooms did have hatches between them to make two-room suites. But no compartments would be bigger than two rooms. This applied to the hotel lounge as well, with two pods that would contain a very small restaurant and a small bar. Most meals were expected to be delivered to the sleeping rooms by room service personnel. There was to be a larger gathering space in the circular hub structure in the center, which was also where the passenger Shrikes would dock and transfer passengers and cargo. Large oxygen tanks, water tanks, and environmental control equipment would also be in this hub area. Adam and his people were now going to observe the facility where the peripheral rooms would be manufactured.

This would be done via 3-D printing in a fashion long envisioned, used, and endorsed by the space community. It had actually already been used by NASA to provide a new tubular-shaped compartment for the ISS. Each pod was really just a massively updated use of the old "Bounce House" inflatable toys that were rented out by children's party companies for decades. Shoot, they're still used today. A flexible inner body was fabricated with attachment points for various preplanned equipment directly built into the structure, both inside and outside. The inside attachment points were meant for everything from interior bulkheads, cables, pipes, vents, and even furniture. The outside points would accommodate external sheeting for protection of the interior, antennae, lights, support struts, and so forth. The outer cover sheeting would have self-sealing capabilities, radiation shielding, and extra tensile strength. Everything needed for construction would be parceled into discreet packages that could easily fit in the cargo bay of any Shrike. Once in the appropriate orbit and parked at the construction site, the cargo bay doors would open, and the inflatable cylinder/room would be positioned outside using the robotic arm. A large tank of pressurized gaseous oxygen would also make the trip with each module and would then be connected to the uninflated cylinder. Once outside, the cylinder would be pressurized, and the sausage-shaped room would inflate to its full size. Each would measure 50 meters long and 10 meters wide. There would be docking ports on each end, and the robot arm would be used to connect the new room to those already in place. It was calculated that an elbow-type adapting module could fit in the same Shrike as the pod being installed. There would be thirty rooms total, with a larger docking port to the outside present on the adapting module in every fifth room, for a total of six ports. These would allow access to appropriately equipped Shrinks to bring in supplies to the interior or to extract passengers from the ring in a hurry if there was ever an emergency.

Exteriorly, rails would quickly be installed to the outside of the pods to allow the newly constructed robots to fasten on and move freely over the exterior surfaces. The robots would be put to work, attaching the exterior protective (and decorative) other panels and equipment. The entire process had been choreographed well in advance and was a masterpiece of planning. This was only made possible via supercomputing. A cargo manifest had already been generated for every Shrike mission, complete with the layout of every single piece of cargo to be loaded aboard. These also included the weight of each and the amount of fuel necessary to carry on every mission.

That morning at 7 a.m., Adam alerted Azaan Nazari and Gregor Levinski that he had arranged meetings with appropriate designers for the project to review blueprints for the various systems to be installed in later missions. Iza Mazurkiewicz would also be sent to observe the robots to be used in station assembly. All of these designs were patented, and they would only be shown plans that were relevant in the assembly phase. But Adam knew the trained eyes he was sending could gain enormous insight into the general principles involved in their designs. On further reflection, he impulse e-mailed Mary Dupree and asked if she could come to the flight line right away since they would be observing the manufacture of an individual module at ten o'clock that day in Phoenix. The remainder of the team was advised to develop summaries to present, giving a thumbnail sketch of those aspects of their specialties that might be highly pertinent to the goals he had presented yesterday. These should be ready by next week's meeting. Those given travel orders should prepare to fly to their respective destinations tomorrow. On Monday next week, the Task force would reassemble to compare notes.

The trip to Phoenix was not particularly inspirational, but it was really fun to watch. Adam, Manjeet, and Mattie were familiar enough with Uplift manufacturing processes to deduce that the 3D printing they observed in Phoenix was comparable to any equipment or facilities Uplift already had in place. One thing they hadn't thought of, but probably eventually would have, was the arrangement of solar panels on the exterior of each of the individual pods. These would be attached after the external shells were put in place. All of these tasks were expected to be completed by robots. Mary Dupree did take especial interest in the composite materials used to make each pod flexible enough to be inflated in orbit. Cables and vents could also be printed into the very skins of the modules as they were made, as well as a vast complex of attachment points and interfaces on both the interior and exterior surfaces. The composites were designed to significantly stiffen after a few days' exposure to the deep cold of space, making them quite solid. Hard panels made of thick steel alloy would be attached to the exteriors to make yet another layer of "shell" for the pods. These would be added on subsequent Shrike missions. When this was completed, tanks would be brought into the interior with pressurized composite insulation material that would be injected through special ports to provide insulation from the cold exterior temperatures between the walls of the pods and the exterior shell. This would protect the interior from radiation as well. The foam would also make all of the pods self-sealing in case of micrometeor impacts. The old ISS had used some similar technology, but not as thickly applied.

Actually, the biggest challenge in construction appeared to be the central hub portion of the hotel. This part of the structure would be far too large to fit into a Shrike's cargo bay. It would also be far too massive to launch fully assembled into space on even the largest available rockets currently available, or even those currently in the design phase. The solution was to start by sending up unfolding frame structures which could be joined together with welding capable robotics. The extendable robotic arms of several Shrikes at one time would need to be closely choreographed to work in unison to allow for this assembly. Adam confirmed that if computer assisted station keeping and arm manipulation were employed on the Shrikes, this methodology should be possible. Range finding radar was standard on all Shrikes, so the computer brain should be able to keep the ships in whatever position requested. Juni would be busy for the entire build process working out these algorithms. He was glad she was finally fully employed with the company, as she was the ablest person for this type of work he had ever seen. Once the frame was assembled, a huge circular disc with an oblong rectangular tunnel through its entire width, access ports, interior bulkheads, storage tanks, and a myriad of electronics, pumps, communication equipment, and much else would need to be fitted within it. The robots would then install the exterior skin and the tubular struts to the circular room/pod section. There were plans about how to distribute cargo and personnel between flights for this entire process. With the use of no less than 90 Shrikes making ten flights or more apiece, the process should take about three months. The individual flights could take as much as one week apiece, usually with three ships together in orbit at the same time, and in different areas of the structure. With the parent company paying for all equipment, fuel, and personnel salaries, Uplift still stood to make a profit of about 15 billion dollars. This was not especially profitable, but the experience gained in massive zero-G construction projects, profits would surely begin a rapid uphill trend. In fact, most of the basic skills required for these types of flights were somewhat old hat for most Shrike crews. It was the scale that was daunting. Everyone employed would be working long hours for weeks at a time until the project was complete. But those hired for this job would see great increases in their personal income. They would also gain some rather impressive content for their resumes for future jobs they might want later anywhere in the Aerospace sector. It was estimated that the project, once started, would take about five months.

The Monday following the Phoenix site visit, Adam's task force met as scheduled. The minutes were recorded, as all future meetings would be from now on. From this point, all content presented here will be based on these records, as has already been stated. Adam again asked what ideas each member had compiled for how their particular fields could be employed in designing a self-sustaining space station in the quickest possible time. This last bit was something of a surprise to everyone, as nothing had been mentioned previously about any time frame being stipulated. Nevertheless, everyone had some input to contribute. Here's what they each had to say:

Juni stated that a supercomputer which could be used in navigation, system monitoring and control, communications, and data analysis could conceivably be assembled in packages that would fit in five or six Shrike missions to be installed for the theorized project. This main computer would take a constant power supply consistent with an output of eight hundred kilowatt hours per day to operate properly, not counting cooling and atmospheric moisture control, which were essential. Adam threw her a curve ball and asked if a general information bank could be added which could hold a library of information rivalling the internet as it now stood. This should hopefully contain the equivalent of the contents of the Library of Congress from this current year. She was not entirely pleased to hear that expectation, she said. That was an entirely new question that would take more time to answer, not to mention more power, space, and money. She did commit to saying that something smaller, with more precise parameters might be possible. Adam said that, no, he expected the any computer aboard his imaginary station would need nearly the entire content of human knowledge, or at least very nearly that much before he would consider the project as viable, Juni confirmed that she understood this request.

Manjeet was next. He stated that he realized that he needed to verify whether or not this hypothetical station would be operated in a zero-G or an artificial gravity situation on board. No one else had given any thought to anything but a no gravity environment, but Adam was glad the point was raised. Prior to this, much research had been done on using rotating surfaces as a means of simulating gravity. A structure, like the one in the hotel about to be built, could be rotated so that the outer surfaces of the wheel would be subjected to centrifugal forces that would press those inside towards the periphery of the wheel in a feeble mimicry of gravity. The hotel would be able to generate about 1/6 of a G without making its passengers dizzy and ill for their entire stay. That feeling of spinning would be the primary sensation experienced while on board the station. More G force could be generating by rotating faster, but it would be impossible to stand up, or to keep one's inner ear from utterly panicking. These principles were well known. Research had at least shown that only a very large circle rotated at a much lower speed could be habitable in the long term. "How big a circle are you talking about?" Adam asked, knowing that Manjeet would have thought this through quite thoroughly long before coming to this meeting. "About 3000 meter's circumference, "he said. If we assemble a ring of 20-meter-long modules and did two rings as we are doing with our current project, that's about 300 modules. Of course, we can manipulate sizes and groupings of these "rooms" in any way we'd like, but that's the size of things. We could experiment with the rotation rate when we're up there to see what works best. There would always be that sense of motion, though, like sailors at sea. We, or whoever would get used to it eventually, but it's still better than zero-G in countless ways." Most everyone in the room groaned and sighed. 300 modules seemed like a lot, but Adam was unfazed. "Let's see how things go with this current build. Actually, it's not as bad as I expected."

Gregor Levinski was next up. His report was somewhat shorter but had revelations all its own that provided food for thought, or perhaps, water for thought. "I think appropriate technology is pretty well established as adequate in terms of heat distribution, air circulation, sanitation and atmospheric maintenance," he said. "Current orbiting space platforms make good use of converting water, H2O, into Hydrogen gas and breathing Oxygen. We use it ourselves to make our own fuel for the Shrikes. The electrolytic process is pretty efficient, but not perfectly so. We wind up delivering a lot of liquid oxygen to orbit to make up for leakage and so forth on current stations. We could probably do better, but you say we're lifting 150 modules for habitation on this theoretical project? Let's say 20 meters long times maybe 8 meters wide? Let's see, that's about 160 cubic meters of water per module. Well, even if we only use two-thirds of the modules, that's way more water that has ever been aboard one spacecraft before. We should be able to shift between Hydrogen, Oxygen, and water pretty freely, and for whatever period of time needed, at least in our lifetimes."

"And Adam can have all the fish he wants!" Mattie injected.

"Yeah, I was thinking that," Adam agreed. "That reminds me, I really have to hire someone to address food and nutrition. I assume we can use water for growing crops, fruit trees, and various other nutritional applications. What about human waste issues, Gregor?"

"Pretty standard, I think. We should certainly go ahead and recycle the water waste. Every drop will be precious no matter how much we bring aboard. Solid waste should be sanitized and used for compost for these plants you're envisioning. A waste treatment plant should be designed to occupy one of these modules, or better still, it can occupy this central hub you're envisioning, since it will be more remote from the occupied areas. A lot of pumps and tanks will be needed. Not to mention conduits. The bigger the crew, the more space we'll need."

"Oddly enough, we'll be installing systems just like you've described on this hotel project for some type of hypothetical future structure of our own," Adam confirmed. Gregor stated that he was aware of this as well.

"In that case, I also think we will need to carry both salt and fresh water for any on board aquatic life forms we carry along, with the capability to refresh both types," he hypothesized. "That should be doable. Plus, the water tanks will need to be temperature controlled. That's a lot of heat to generate on top of what we warm blooded creatures already require."

"Understood", Adam said curtly. "OK Awamila. If you were aboard this station, what would you like to have at your disposal?"

The astrophysicist from Pakistan seemed somewhat amused by such a question. She closed her eyes and began to speculate. She had had no notes from which to speak, but apparently didn't require them. "Well, a state-of-the-art high-powered telescope, for starters," she began. "I mean, the Hubble scope and ground-based scopes were all well and good. If we were to just stay in orbit around the Earth, nothing we could lift would be of any more advantage than that. There'd still be a problem with light pollution though. Being free from the Earth's albedo would be a real benefit, or any other's planet's albedo either! It'd be a crime not to engage in a lot of astronomy work! I guess we should just get the best scope we can. It would be pretty expensive, I'm afraid. Let's see, oh yes, sensors to monitor for things like ambient radiation and subatomic particles in the space we traverse. We could see what might be floating around out there that we can't see or detect from Earth or places our satellites have been already. That would be nice. Oh yeah, I was thinking about this in my sleep last night. I suppose we ought to have some sort of long-range radar to detect any navigational hazards that might be out there waiting for us to collide with."

Manjeet spoke up. "Oh, you must mean if we actually wind-up leaving Earth orbit. Ground control could certainly alert us if we had any potential collisions in near Earth vicinities."

"Well, obviously," Awamila answered. "I'm trying to think more expansively. That's what Adam asked us to do. I would personally advise that we try not to traverse the asteroid belt. Remember "The Empire Strike Back!" But if you do choose to cross it, radar will be an absolute necessity. Actually, I think I'll be staying at home if you go anywhere near the belt. Now Juni, you said we'll have computer support for navigation? That will be absolutely essential."

"Yes, and computational support for any astrophysics calculations you might require," Juni affirmed.

"Well, that's a good start!" Adam interrupted. He pressed forward. "Azaan? What have you been thinking about?"

The large and sturdy professor from Tanzania was looking quite introspective in his expressions thus far. "It all depends on what you are expecting this very large spacecraft to be doing when you're aboard," he said. "If we're only in Earth orbit, we'll basically need just those reaction control jets to change our orientation in small degrees from time to time. I think we're all kind of down on using Nitrogen Tetroxide tanks for propellant on those, am I right?"

Adam took no time at all to state his agreement to avoid that particular propellant.

"OK, I feel sure we'll have plenty of plain old Hydrogen on board for that purpose." Azaan continued. "Now, if you really want to consider leaving the neighborhood for a while, we have a lot of options. All of these have been considered, and some have already been used in a very limited way by NASA and others as well. Larger scale motors would require a lot of research, and an awful lot of money too."

"What do you have in mind?" Adam sighed. Everyone knew he had already been thinking about this topic too.

"Well, there's ion drive," he began. "This would require us to acquire large amounts of Argon gas, which would be ionized and released slowly on a continuous basis. It would take time to build up speed, but we could keep accelerating for years and eventually be going pretty damned fast. It's slowing down that would be difficult. That would take a long time as well. A few small probes have used this, but for very short distances, in relative terms."

"Gotcha," Adam said. "Go on".

"Well, there are those much-ballyhooed atomic engines", Azaan continued. No one's ever built one big enough for our purposes. The theory goes that one of several types of plasmas could be generated and shot out the back of the engines with extremely high velocity and heat, and for quite extensive periods of time. When we needed to slow down, we'd just swing around and fire in the opposite direction. Nuclear power would also be a convenient method for generating heat and electricity for the entire ship. The more fuel we carry, the longer we could stay out. Several lifetimes, I suppose."

"Yes, I think a lot of us have been considering that," Adam said, rather softly. "Trouble is, the FAA is reluctant to agree to even sending very small amounts of radioactive material off the planet. Even a couple of grams to power a satellite makes them nervous. Even then, they use that stuff only for deep space probes, not large orbiting stations. We'd need multiple kilograms of Uranium or Plutonium for the type of reactor you're envisioning. Actually, I've got a nuclear engineer named Jeter Hamm out of Norway that I'm thinking about hiring. Someone should be available to help us with these kinds of questions. I don't know what he could actually do, though, considering all the regulations and international treaties we'd have to overcome."

"Well, you did ask for other alternatives, Adam," Azaan said. "Of course, there are other options like solar sails, and even been some actual research on Star Trek style Impulse Engines that people are looking at. Some of the really out-there researchers think that Warp Engines could actually be a possibility. I wouldn't want to dump any of my money into that myself."

"Well, as far as I'm concerned, neither would I. Remember I said to keep things as real as possible, unless you happen to find something that your best judgment can support," Adam reiterated. "But why don't you try investigating all the other things you mentioned. We can pay for that. Maybe try drawing up some schematics for basic prototypes for all of them. If you feel it's feasible to build any of them, we'll see what we can do."

"Um, OK. Give me some time," Azaan agreed. "I really have no idea how long it might take. Then I suppose we'll use trial and error to figure out the best one to use?'

"Who said we only need to use one?" Adam said. This seemed somewhat surprising to everyone for just a moment before this assertion suddenly made sense. Adam was by now understood that he truly needed to think expansively at all times. It was slightly uncomfortable for the others to realize he was expecting the same from them. The meeting moved on. Peter Hosokawa was the last to be called. But before he was, Adam called a break for lunch. He had had the meeting catered by the cafeteria people. Every got up and stretched, went to freshen up, and then came back to eat at the conference table. The meeting resumed precisely at 1 p.m., even though not everyone had finished eating. The last segment involved Peter's findings regarding medical care on the project, and it took several hours more just to cover his concerns.

"Well, Peter," Adam started. "What's been going through your mind?"

"To start with, I'm thinking you should really address me as Doctor Hosokawa." This seemed like a bad way to start. Adam, as a self-proclaimed team builder, was hoping to cultivate more of a feeling of equality among the members of his task force. This request just rankled a bit. Adam countered as calmly as he could.

"Well, if we do so, then for equality's sake, you should address each of us by our titles as well. I'm the only one here without a PhD. Perhaps you could call me "Mr. Chairman," and everyone else as "Doctor." Would that satisfy you?"

"Well, not really," Hosokawa replied, quite seriously. "I'm the only one here who's a medical Doctor, and I demand the respect that that entails. All of you are simply engineers, peers. Calling all of you Doctor would just be redundant. I really don't have any peers here, in the professional sense." This assertion was met with barely perceptible recorded groans and hisses of a kind that had not been heard in any previous meeting. It was the first truly non-constructive statement that anyone had made at one of these conclaves.

"I begin to see why your former employers felt obliged to let you go," Adam said, in a rather lamenting tone. "In any team-oriented group, a sense of superiority of some members over others usually just leads to tension and resentment. That sort of atmosphere won't simply work here. We're going to be locked up together as a group for some time, in theory at least, and an artificially tiered system just won't cut it. If you've come here for the purpose of self-aggrandizement, perhaps you should look somewhere else. Private practice perhaps? Otherwise, Peter, present whatever ideas you might have for us. Now would be a good time."

"I wasn't 'let go,'" was the response. "I chose to leave, mainly because of the arrogance of my employers. Alright, I'll present. But I reserve the privilege of addressing this issue later. I think you're overreacting." Adam grunted his assent.

"Very well," Peter continued. "Now since you're suggesting that anyone manning this station must be completely cut off from outside support for extended periods, right? Then from a humanitarian standpoint, there needs to be comprehensive medical care available to everyone aboard around the clock. Think about that. It's easy to say but challenging to provide in the real world. A general practitioner is only the starting point. A GP should be on hand to attend to all the common diseases to which humans are forever susceptible. That being said, I suggest that everyone start out being immunized for the usual contagious diseases as soon as possible. Tetanus, Pertussis, Mumps, and all the usual vaccinations have been ordered for all of you by the Uplift clinic physicians already. You'll all need to have boosters available on board before we consider going anywhere. Fortunately, there will be no exposure to pathogens from the public once on board, just what we bring with us. We should have plenty of vaccine doses aboard anyway though, for any crew additions that may, um, occur while we're isolated up there. I'll elaborate on that thought in a moment. Now, treatments for the natural diseases of aging must also be available, from Alzheimer's to cancer, to depression, and so on and so on. This will require long-term availability for a vast myriad of medications, and probably a pharmacist to manage them. What about surgery? Emergency or otherwise, we should have both a thoracic and general surgeon aboard. I'm glad that there will be at least a moderate gravity field on this vessel, because there can be a good deal of fluid released with these procedures that will need to be contained. An anesthesiologist would be helpful, but most surgeons can be trained to do this. Another problem with surgery is the need for cauterization of tissue and blood vessels to prevent extensive bleeding. This involves use of electric current applied to a probe. This can't really occur in a pure oxygen environment. (Gasps of OMG are heard in the background. No one had considered this). Speaking of large fluid spillage, how do we cope with childbearing? An OBGYN must be aboard for these concerns. Plus, birthing suite capabilities. Again, I'm glad there will be some sort of artificial gravity. It's hard enough to live in zero-G for those who enter it as adults. General body deterioration begins rapidly enough for those people. I can't see infants developing properly without some sort of simulated gravity. It would be abominable to even think of inflicting that on a child." From the tone of the other participants, it was clear that none of them had done much thinking about these sorts of topics.

"That's just the beginning," Peter went on. "What about diagnostic equipment, CT scanners, x-ray machines, and the like? These are heavy and use a lot of power. Lab and measuring equipment will be absolutely essential as well. A large medical library should be loaded up in Suni's computer to guide our doctors, whoever they may be. Don't forget chemotherapy agents and lasers for tumor treatment and other anticancer agents. That's just the medical side. What about dentistry? Ophthalmology for vision problems, glasses, cataracts? Hearing aids? Artificial joints and limbs? You see, these are all huge concerns!"

"I'll admit I hadn't given this topic the full attention that I should have," Adam said somberly. "We will obviously need to choose our crew carefully. It's hard to imagine finding personnel that can cover all these diverse disciplines with the fewest number of practitioners. Not to mention obtaining the equipment and supplies necessary. This will take up a large part of our budget. But it's not an option we can just forget and hope for the best, now is it? We need to be fully prepared, or else no one will ever want to participate in our little project. Would any sane human being?" No one challenged this assertion. "I'm always reading history," Adam said. "Early explorers got on old wooden ships and stayed there for months on end just to try and bring the far reaches of the world closer together. But a lot of people died in the process. We won't be getting reinforcements on our imaginary mission unless we make them ourselves. Then we'll have to wait for them to grow up before they're any use. We'd better be sure to stay as healthy as we can for as long as we can." Time was running short, and other small talk had taken up time that isn't covered in this synopsis. Adam eventually concluded the gathering.

"Well, alright everyone. I'd say we can go ahead and dismiss. If anyone needs to leave, go ahead. Feel free to talk amongst yourselves though if that serves your needs better." The group lingered for another hour or so, making inquiries into things that were discussed that happened to interest them. Or about how their various fields of interest might intermingle for mutual support. It was obvious to everyone that they easily might have committed themselves to an engineering dead-end of cosmic proportions. There must have been a lot of self-doubt and skepticism in the group about this time. But Adam, ever optimistic, had faith that he had indeed chosen the right people for his off-the-wall project. And his faith would only increase with every passing day and every passing week. The example of the Infinite Horizons construction project they would soon commence seemed especially important to him as a means to prove the feasibility of his ideas. This build would start in a matter of weeks. Even so, this meeting served everyone as an example of this team's ability to identify problems and to apply their collective minds towards finding a solution. It also identified some other potential problems that would likely be ongoing. Engineers could solve mechanical and equipment problems fairly rapidly with computer assistance. But successfully meshing together a group of human beings to work for an audacious common purpose might not necessarily be quite so straightforward. Time would undoubtedly tell.

Chapter 8
Let's Build a Space Hotel

Construction of the Endless Horizons Orbital Hotel project began precisely on schedule. A massive storage facility had been constructed quite quickly at Uplift, just behind the immediate structures lining the flightline on the north side. It was climate controlled, of course, but it was not entirely a "Clean Room" environment. Most of the components stored there did not require anything but a dust-free environment. The inflatable modules' interiors were already sealed off from the inside from contaminants. Such sterility was also unnecessary for the large structural framework components, which would generally telescope out into their final shapes and configurations. Again, though, dust was to be avoided at all costs to keep the works from gumming up. As a rule, the Shrikes were backed into the facility and the huge doors closed behind them when loading cargos. This had generally been the case for all other Uplift flights as well. This first mission would contain the structural components of the central disc, about one quarter of the internal framework. The first two construction robots would also be deployed right away. Adam was to be the primary pilot of the very first flight, with Juni in the second chair. Manjeet was Flight Engineer. Iza Mazurkiewicz would be an observer for deployment of the robots, and the hotel company would be sending their own robotics specialist along as well. About six hours after this launch, a second Shrike with the second quarter of the disc would be launched with Mattie at the controls. Nathaniel Floatingfeather was copilot, a duty he sometimes performed, and Gregor Levinski was engineer. All further flights would include personnel both from the hotel company and from the "Sustained

Mission" task force. The idea was to include the task force members intensive spaceflight training and eventual certification.

Juni Sato had had time to review and often augment the computer protocols which would be used to automate much of the construction process. This included use of the robot arm, radar assistance, and navigation. The robots were another story entirely. The Horizons company had independent control of them during the construction process. All the Shrike crews could do was observe, at least initially. Iza would be working with the robot's manufacturers on an ongoing basis and tried to learn what she could of how they were being used. Most of those technical details were trade secrets, and it would be too much to ask for them to provide a step-by-step narrative of the details while the build was actually going on. All that mattered was that everything seemed to be going pretty much as planned, and it was a very complicated plan. Use of supercomputing, along with the remarkable agility of the Shrikes and their associated equipment had made it a smooth and very predictable process. In two short weeks, the donut frame of the hub and docking bay was assembled. The frame network was designed to lock in place with internal latches when the components were inserted together. The external plates were then snapped into place to make a tight seal. Larger and heavier components to build the docking bay required Shrikes to manipulate them around for assembly when they arrived after being lifted into orbit with heavier conventional rockets. These rendezvous all were accomplished without incident, and the cargos were all utilized seamlessly with the Shrike's remarkably dexterous arms. Heavy tanks and equipment were installed into the central disc interior, and internal bulkhead plates were successfully fitted into their proper positions to make interior rooms and compartments. Finally, the exterior plates were placed. Two robots were offloaded into the interior, as were four more brought up on later flights to be used outside. These operated by moving along various rails and mounts placed both inside and outside the structures using analog arms and legs to which various tools could be attached. Their feet could grasp onto rails and handholds in the same way as many primates can. They didn't "jet around" with autonomous rocket packs or anything of that sort. They needed to be anchored to some solid object using their hands or feet before they could perform any useful work. They had drills, impact wrenches, and even equipment for arc welding at their disposal. On no occasion during the entire build was it necessary for any humans to do extravehicular tasks. All operations could be managed from within the Shrikes by human operators. Crews were needed mainly just for monitoring automated operations. On a few occasions, manual control of the robotic arms was needed to make minute corrections. This required a bit more finesse than any robot could provide.

Construction of the four radiating spokes, and then the circular tube structures required a further four weeks. The prescribed flight schedules were adhered to quite precisely and without difficulty. This feat was a true testament to the value of advanced planning using both human and computer input. The eternal heroes were the Shrike spacecraft themselves, as well as the crews that operated them. Frances Thorne had designed some very solid and versatile spacecraft. Adam only wondered why he hadn't really utilized them more. He would have done it differently if he were the one in charge, he thought to himself. But then, maybe the feisty birds needed to prove themselves first. They were certainly proving themselves now, and it made him enormously proud. Principle construction of the primary hotel structures were completed about on schedule in six weeks, with as many as six Shrikes present at the site at any one time. The next phase of the build would involve actually docking the Shrikes onto the station and unloading cargo through the docking ports into the interior of the structure. And there would be massive amounts of cargo. Once there was a pressurized interior atmosphere, humans could go in and assist the robots in assembling all the hotel accommodations. Uplift personnel of all types and specialties were taken aloft in greater numbers than ever before. After the first two weeks, the crews could begin sleeping on the station, and not on the Shrikes. This enabled more people to be orbited with each mission. Seating on the Shrikes was only needed for launches and landings. No one even had to sleep in the cockpit. The birds were monitored from the ground and could be left without crews inside. Berths and food were provided throughout the hotel spaces, sometimes as many as fifty workers were in orbit at a time. This was a new world record and then some. A monumental amount of wiring, environmental control equipment, decorations, supplies, communication equipment, navigation equipment, and all the other necessary hardware was orbited and assembled. Adam Thorne thought about staying up permanently to supervise, but since he was rotated to pilot a mission every other day anyway, he decided that that was unnecessary.

The project moved forward with such precision that even the designers of that precision could scarcely believe it. The astonishing thing was that everyone soon realized that it would have been possible to have done a project like this as soon as Shrikes had been invented to do it. But humanity, with all its ingenuity and resources, had been too timid and skittish to ever try it. The station was every bit as luxurious, though nowhere near as large, as a high-end hotel on the surface of the Earth. The only thing missing was a ballroom. That would be unnecessary though, as one could feel like they were dancing simply by jumping off the floor. Of course, guests would be in Earth orbit and had only enough gravity to touch the floor when standing perfectly still. A moderately healthy person could at least move around the place when they needed to. The construction people were soon experienced at living in these conditions and eventually began to actually enjoy their work. It was everyone's hope that the necessarily wealthy potential guests would feel the same. Only time would tell.

When the project was done, Adam was quick to congratulate the employees involved, companywide, with praise, promotions, and healthy bonuses. He decided to put a three-month hold on meetings of his special task force. All other Uplift operations would of course continue, as there was no real need to suspend them. This also gave him and everyone else on the task force an opportunity to take some leave. Three months seemed like an appropriate amount of time, and that is what he approved. It was the first vacation of any sort he had allowed himself since taking over for his father. Like his dad, he absolutely refused to indulge in a jet-set sort of lifestyle, or at least he mostly refused. He certainly made it a point to fly under the radar of the media. He generally would rather fade into the background than sit around at Uplift waiting to be interviewed by the press. Even so, in the future months, he would find himself testifying before Congress and other Federal committees. Some of that was highly publicized, and that kind of stuff had been a longstanding part of life for all American billionaires and financial giants. But for the two-man Thorne family, being highly visible was basically considered a sin that should be avoided as much as possible. This was not because they had anything to hide. It was more because Frances and Adam were always too focused on working to allow for any distractions. What the world thought of them had never mattered to them one iota. Public opinion would be influenced by the results of their work, not how charismatic they appeared to be. During this break however, and just for this little while, Adam and his closest associates needed a small interlude to explore themselves for a change. They needed to simply recall that they were mortal human beings with mortal human needs. Most of these recollections about this time come from diary entries transmitted from off the planet by Adam and Mattie themselves a bit farther in the future.

"After we got done building our first big orbital structure," Mattie wrote, "Adam saw fit to grant us all three month's leave to explore our own pursuits. I guess he felt pretty sure we weren't planning to resign and go off on our own. He was right, of course. I know he had no doubts about me because we had been living together for some time by then. Plus, he had known me intimately for a few years before that. He always knew if I were being straight with him or just saying what I thought he wanted to hear. Therefore, he considered me an actual friend that he could rely on. It seems really weird in retrospect. I didn't really know for sure if he was who I thought he might be. But if he was, then he should have been a big playboy type with women clawing down his door all the time. At least he should have been a massive party guy, or some kind of media darling. But in reality, he was the most boring man I'd ever met, a true engineering geek. He was never mean or condescending though, not to anybody. If he disagreed with you, he'd just calmly state why and then hope you'd probably see it his way in the end. He was usually right about that. As for me, I am more of a hothead almost all the time, I wasn't sure we'd be compatible. But he always listened to me, and usually didn't shoot me down. There was no doubt I would fall in love with him, he knew it as well as I did."

"After we moved in together, we were as good as married anyway. The only disagreements we ever had were work related, about how to solve some problem or other. But that was only work, and engineering problems require the work of entire groups anyway. Those are not one individual's responsibility. As far as problems in our relationship, there never really have been any. We always discuss any differences we have and usually are more interested in what's best for the other person as opposed to what's best for ourselves. When I first moved in, I told him how I felt about him after I'd worked for him for a while. He thought I meant I loved him professionally, what a geek!. I put him straight about that pretty quick. I guess that couples that love each other are mutually employed on a personal level as well. Our job goals in the romantic sphere are to give the other person as much love and happiness as we can. In that way, he was working for me as much as I was for him; he just didn't know it. After that stressful period building that infernal hotel, I saw he was a man who could keep his cool, for the most part anyway. He wasn't the kind of man who got snippy with others when he was stressed. I think most men tend to do that, especially with their wives. Not Adam, he was always as kind and respectful to me, even when we were working closely together. That's how I knew he was the one to spend my life with. It also amazed me that I was the only woman he ever thought of in that way. Manjeet confirms this. Despite being one of the richest bachelors on Earth, he never once dated or ever even looked at any other woman but me, or any man for that matter. We'd known each other for, what, four years? It was only then that he allowed himself to express that he truly had those deep feelings for me when I agreed to live with him. Neither of us wanted children at that point, so we did the birth control thing. Neither of us had any living family, except his mother I guess. But she never even called him when his father died. She still never has. The funny thing is that he continued to hold off on wanting children for as long as he did. "Let's see how things stabilize before we create any new humans," he would say. I think I knew what he was implying, and I turned out to be correct. I guess that when we really bonded, our brains kind of fused together."

By the time this work hiatus began, Mattie was the first to really embrace the idea that she had "made it," both as an Engineer and as Adam's true love. Adam was thrilled that Mattie apparently loved him for who he truly was, and not because she wanted his money. That was his mother's legacy. He just wanted someone who could be compatible with him, and that order was tall enough anyway. By some incredibly good fortune, they had met and found they simply loved each other, and that was the only goal either of them ever had. That is not to say Mattie didn't start to use Adam's resources to indulge herself a little. As an engineer, she found her love for the Thorne family cars too much to resist. She didn't start begging Adam to borrow one of the many vehicles in the family fleet; she simply began to treat them as her own personal property. Adam was thrilled about this too. A trainer was hired, and Mattie was found to be an outstanding pupil. She was already a trained astronaut, for heaven's sake. She began taking trips throughout the southwest to race at local tracks with the family racecars as a serious hobby. She took a few trophies but never any first-place finishes. This all started during this hiatus period, and she continued to work it in during her remaining ground time while working at Uplift. Adam's hobbies outside of work were few. These haven't been mentioned before, as he couldn't work his leisure time into the rest of his schedule very often. He usually just spent his time in his home study researching his engineering ideas. Early in life, he did develop a deep interest in history though, especially New Mexico history. He seems to have never lost interest in it either, as you may have noticed. This has always been, and continues to be, a bit of an obscure topic, especially among New Mexicans themselves. The state has assumed all manner of identities and cultures before anyone ever conceived of it being a part of the US. Many of the other members of the United States still have no idea that New Mexico is one of their sisters. This phenomenon surfaces every time a New Mexican travels to, say, New Hampshire. Perhaps a hotel clerk might remind him or her that only US currency will be accepted there. It is both exasperating and hilarious to experience. But New

Mexicans are perfectly content with this odd chasm in the consciousness of their countrymen. Mainly this is because it allows New Mexicans to live more tranquil and unflustered lives than other Americans. When the average Nuevo Mexicano watches the evening news, he is generally perplexed at the goings-on of those outside their borders. Not that there is no crime, poverty, or general strife in New Mexico, there's as much or more of that than there is anywhere else in the country. It's just the general level of hysteria out there past our borders that is so hard to comprehend. New Mexico's temperament just seems more laid back, more rational somehow. Even the residents of the midwestern states, the heartland it is called, always seem more agitated and somewhat frantic about issues that New Mexicans simply toss aside as being trivial. For us, if outsiders think our "Land of Enchantment" is some sort of alien-infested third world nation, then let them think so, and continue to forget we exist. We'll handle things ourselves, thank you.

And the Thornes did handle things, almost everything in fact. Frank, Adam, and their company have been a source of prosperity and lifestyle improvement surpassing anything New Mexico has ever seen before. Poverty by this time was at an all-time low, for New Mexico anyway. Still, scarcity of water has and will continue to be a major problem here. Rainfall is rare despite the reduction in global warming since the phasing out of fossil fuels. New Mexico will remain a desert for the foreseeable future, but it remains eternally hopeful. Even Arizona, though hotter, at least has some respectable rivers within its borders. The Rio Grande of New Mexico doesn't ever seem to have more than a trickle of water running in it until it reaches Texas. Then other Texas sources join with it to make it much more robust. It's all very unfortunate for the residents here. Nevertheless, the underground aquafers in New Mexico seem to be refilling lately, allowing Uplift and other companies to electrolyze the H2O into the H2 and O2 it needs to power themselves. This has allowed the unexpected transformation of New Mexico into one of the greatest gateway to the stars mankind has ever seen. Sorry Texas, sorry Florida. This all has arisen in one of the most remote and ridiculed area on the entire North American continent. Adam realized New Mexico's ostracism very early on in life, and it both fascinated and befuddled him. From the earliest inhabitants who built the cities of Chaco canyon, to the Pueblo tribes, to the wandering nomadic tribes of the prairies, the dwellers of this region have had to adapt to a life with scarce resources. Their societies also managed to devise a near perfect union with both the land and the sky. All subsequent inhabitants of this region since have needed to adapt to the harsh environment as well. Even those first Chaco cultures were as aware of the arrangements of the heavens every bit as much as even the most updated Uplift supercomputers. Nathaniel Floatingfeather has always assumed that some ancestor in his remote past must have been fascinated with flight in some way that earned him his family name. For all we know, this forebear was even better at conquering the mysteries of the sky than he has been. The progression of events in New Mexico from that remote time forward has been a continuous saga no matter what period one examines. Such is the New Mexico. What most intrigued Adam is how New Mexico still is always ignored, or even distained, for reasons he could never understand.

Now, a there are a few things about New Mexico people seem to remember. For the most part, unfortunately these are bad things. Billy the Kid and the Wild West are pretty well ingrained in the public memory. The Sante Fe trail too, maybe. People liked that old TV show Breaking Bad, but that was about a drug dealer. Some new Mexicans remain irritated about that. Details about the Spanish colonization of the area and the exploitation of the indigenous people there is taught in most American schools, although many have tried to downplay or even erase that memory. This erasure is still a goal for many. Also, New Mexico remains one of the more Catholic regions of the country and has more people of Hispanic heritage per capita as well. Any true New Mexican, however, wouldn't change a thing, regardless of their own cultural heritage. To do so would be like Louisiana disowning French culture, no matter where their own family originated from. Try to tell a New Mexican that they can no longer eat a green chile' chicken enchilada and see how quickly you are urged to return to wherever the hell you came from in the first place. The same for refried beans, sopapillas, tamales, and all those other tongue-searing foods. As soon as Adam Thorne took over from his father, he made sure that any flight he was commanding had a healthy supply of New Mexican cuisine was on board for him to eat. This was unnecessary since all the flight crews had made the same demands for years. It had been an option before, but now it was a policy. Chile is addicting stuff.

All this being said, Adam had been interested in the furthering the cause of historical preservation in New Mexico as early as his teen years. He seldom had much time for him to do much about it, but he would try as much as he could. He nudged his father into donating moderate sums of money to various organizations and institutions he heard about that shared his goals. He even dabbled in historical re-enacting, mostly during the summers starting in high school. New Mexico had a small cadre of Civil War reenactors that really drew his interest. He chose to join on the Union side. It surprises most people even now that two battles were fought between the Union and Confederacy in 1862 in New Mexico to decide who would control this far western portion of US territory. Although these battles were comparatively small in comparison to the massive slaughterhouse fights in the more easterly portions of North America, this southwest part of the country was really quite important. Due to the recent gold strikes in the area, not to mention the chance to expand the

Confederacy as far as the Pacific coast of California, the contenders from the South had a chance to accomplish a great deal with the few troops that gathered at the Alamo in Texas, planning to invade the vast New Mexico wastelands. What they hadn't counted on was the veracity of the remaining regular army troops there that hadn't changed the colors of their coats from blue to gray. Even more surprising to modern eyes were the large number of volunteers from the general population of the territory that chose blue. These were young men who had begun their lives as citizens of old Mexico itself. After the Mexican War, their parents had elected to stay where they were, rather than move southward to remain loyal to their original nationalities. Most of the young men who volunteered for the Union had yet to learn much English and had volunteer officer who hadn't either. Despite these supposed liabilities, they showed up by the hundreds to receive uniforms, equipment, and training. They stood up proudly to combat the land-greedy Texans at a place called Valverde, along the Rio Grande south of Socorro near a Union outpost called Fort Craig. They were placed alongside the hardened regular soldiers stationed at the Fort, who were undoubtedly glad to see them. Despite the help, the Confederates won the field that day, but they thankfully failed to capture the fort. Foolishly, the Confederates left them there inside the walls without trying to capture them, and simply continued their march northward in the direction of Colorado and the goldfields that lay in that direction. To their dismay, they found the natives they encountered on their northern advance unenthusiastic in their reception of their former countrymen. The Rebels had assumed that the civil populations would welcome them as liberators and assist them in obtaining food and supplies. These commodities which were rare enough in New Mexico to begin with. So no, the Nuevo Mexicanos would accept no Confederate money for their goods, and they hid away any food or livestock that the Confederates would have confiscated if they had the chance. The Army supply depots in Albuquerque and Sante Fe were tempting future targets, but the Union soldiers burned these, and hurried northward to take refuge at Fort Union, north of Sante Fe. The southern army soon found their own supplies dwindling away as the trudged further and further from their Texas homes. They also discovered that New Mexicans were a wily and resourceful bunch. Having lived their entire lives in a remote and unforgiving land, they knew, and still know, how to do more with less.

Back at Valverde, for example, a recent Irish immigrant named Paddy Graydon had resented the idea of a Texan invasion of his new home and pulled together some acquaintances to form what Graydon called his "Independent Spy Company." This bunch of semi-outcasts, like most New Mexicans, knew their territory quite well. They to it upon themselves to scout out the Texan movements and report them to Union commanders. They would even wander right into Rebel camps on the pretext of selling apples or other produce, then they would scamper back to Union headquarters and spell out what they had seen in great detail. Graydon was of especially great service at Valverde. He concocted an idea where he would load up two mules with heavy packs of explosives and bring them to the vicinity of the Confederate camp. He would then light the fuses on the packs, slap the mules on their butts, driving them into the camp as the spies turned and rode off as fast as they could. This they did, but the poor mules figured out what was up, and turned to follow their rascally masters. Graydon and company had horses that were faster, thank heavens, and pulled far ahead before the mules were vaporized. No real damage was done to the Confederate camp, but it did frighten all of the horses and mules tied up there. Dozens of these animals panicked and broke free of their ties, making for the Rio Grande River a few miles to the west. This enabled the union side to capture them, thus condemning many a poor southern boy to walk for the remainder of the campaign.

The supplies remaining on the wagons had to be moved to the wagons that still had beasts to pull them. This complicated things even further and reminded the invaders of just who they were dealing with.

The Union officers were in fact quite a formidable group. The Confederate commander, one Henry Hopkins Sibley, had been an old desert fox himself before leaving Union's ranks to head east to join the rebels. He had invented desert-specific gear for the army, including a teepee-shaped conical tent, known just as a Sibley tent, that could hold a dozen men sleeping like spokes on a wheel. In the winter, a cone shaped Sibley stove would be placed at the center of the spokes to keep everyone warm. But Sibley had developed a bad drinking habit in his lonely desert years, and this habit persisted badly as he led his troops north. His opposite number was Colonel Edward Canby, Sibley's old friend, who was just as wizened and desert savvy, but much more sober. Kit Carson was there too, and led an entire regiment of volunteers, many of them old trappers and mountain men from Taos and other rugged places in northern New Mexico who didn't scare worth a damn. Carson was a living legend, much like Davey Crockett at the Alamo. He was famed for ferocity in fighting various opponent with hellish ferocity alongside the explorer turned Presidential candidate John Fremont. Fremont was an Army officer when the two first met, and Carson thereby found his way into the Army when the Civil War broke out. Carson was by now a full colonel himself. He stayed and guarded the southern part of New Mexico as the Confederates progressed northward. There, again, they found all useful supplies in both Albuquerque and Sante Fe burned to ash, and therefore probed northwards through Apache Canyon en route to Fort Union, which is why they had come all this way in the first place. There to receive them were a fair number of regular army troops and a contingent of Colorado volunteers who had joined up and trained in a hurry for this very purpose, just as the new Mexicans did. They had the added challenge of having to march as fast as they could to arrive here through a deep freeze and blinding blizzard, mostly on foot. They were able to rest and resupply only briefly at Fort Union before surprising the rebels in this narrow canyon with high steep walls. This was just as well, as Fort Union was really more of a supply depot than military bastion. It seemed to have been placed at the bottom of a huge bowl in the prairie, and any Confederate commander would have easily surrounded the place with cannon up at top of the bowl, allowing him to shoot the poor helpless soldiers down below. Union commanders appear to have realized this in time and sallied forth out of their bowl to surprise the rebels instead. They met them north of Apache Canyon at a place called Glorieta Pass.

It was therefore a confined and really intense battle. It was fought in a rocky forest area, not considered an ideal place to fight in the precise, Napoleonic way the tactics of the time prescribed. The Confederate had already been tested once in battle, a claim that these Colorado volunteers had not. In point of fact, the Confederates won the field, as they had at Valverde. But once again, unconventional tactics had injured the invaders severely and unexpectedly. While scouting for a way to circle around the Confederates and thus strike them from behind, Colorado Major John Chivington stumbled upon the entire southern supply train still concealed back at Apache canyon. The union troops descended the steep valley walls and attacked the rear guard. They killed or set lose all the remaining horses there and burned all the supplies that the rebels still had. Though they had won the day, and could have advanced all the way to Fort Union, Sibley knew the jig was up, and decided to retreat. Having no other safe haven, the retreat would go all the way from northern New Mexico back to the starting point at San Antonio Texas. Canby, having no way to feed and care for the hundreds of prisoners he would have taken. He simply let them go, taking up the rear to make sure they kept on going. The rebels had no supplies, even food or water, and didn't know where such things could be found in this strange barren land. It was what would now be referred to as a death march. Bands of the poor Texans were attacked by marauding horse-borne indigenous warriors. Far fewer Texans returned to Texas than had left there in late 1861. They never tried to return.

None of this has the slightest bearing on the overall story we are telling here. It has merely been included because it was the sort of thing that Adam could be very passionate about outside of Spaceflight. His father indulged him enough that by the time Adam was an adult, he had amassed a number of actual and reproduction military weapons and clothing stretching from Spanish Colonial times to World War Two. This was paradoxical, as both of them were staunch pacifists in every way. Reenacting was certainly a frowned upon activity by this by the 2020s or so. The majority of the public seemed to feel that the errors of America in the past, discrimination, and internal discord mainly, ought not to be "celebrated" or "glorified." Perhaps they should even have been forgotten entirely. This always distressed Adam in particular. He felt instead that such mistakes could only be avoided in the future if the past was remembered and reflected upon. Who, he felt, could look at these outrageous events and not decide it would be a horrible mistake to repeat them? But the first step would be to truly and correctly remember what actually happened in days past, not try to forget that it ever did. Besides, would people have ever stood up to defend the best traits mankind had to offer if they thought they would eventually be forgotten? NO! That's what Adam would say when he drove off for a reenactment. He kept on going to these events all through high school and college as often as possible. Many of us Uplift people went with him. We could at least afford the hobby, which really can be kind of expensive. It was even more so for those folks who bought things like cannon or cavalry horses. Adam was the most passionate about it and loved the infantry group, learning things like how to march or fire a musket. For the rest of us though, we just thought it was more fun than something like, say, laser tag. None of Adam's closest friends, like Manjeet or Mattie, ever got into it. The point of all this is that Adam was and is a New Mexican right down to his bones. He had New Mexican heritage in his DNA, as most of us here at Uplift do. He and Mattie often conversed in Spanish, or the New Mexican dialect of it, when not conversing with others.

During this hiatus period, Adam did go to a reenactment of the Battle of Valverde. The battlefield was only a few dozen miles directly north of the Uplift complex. It was located on state property that had formerly been on a large ranch belonging to the late media mogul Ted Turner. Turner was a civil war history lover himself but had donated most of the huge ranch the battlefield was on to help establish a wildlife preservation area. When Adam became a mogul too, he began to contribute money towards buying the land himself. He made the stipulation that the actual battle site would be preserved separately from the preserve. The site had been infiltrated with invasive salt cedar trees that became so thick you couldn't even walk between them. He volunteered to remove the trees and replace them with types of vegetation that would have been present in the 1860's. The course of the Rio Grande, which had changed when Elephant Butte Dam was constructed, would be re-dredged restore it to its original position. In exchange, a reenactment would be allowed to be held there every February, the proper time of year, and the event could be open to the public. Care would be taken to not cause undue environmental damage to the area, or to nearby Fort Craig, which was in ruins in any case, and quite fragile. While he was at it, he decided to donate a few million dollars to help restore and preserve various other New Mexico historic sites. All of this occurred during the three-month hiatus, and it would be the last year he ever participated. While Adam was out playing soldier, Mattie had hauled out a vintage Alpha Romeo to take to a race in California.

The main reason to bring all of this into the narrative is to demonstrate that Adam had spent much of his spare time in life studying human history, US history especially. American political history was equally as important to him as military topics, although he never once considered actually participating in either. He refused to donate a dime to any candidate, and never publicly endorsed a single soul. He never failed to criticize a new policy or position taken up by any politician he thought to be corrupt or harmful to the public interest. This criticism was only expressed among friends, however. Since he never donated to anyone, he would only gripe to people around him if the topic was helpful when making leadership decisions. Occasionally, he would recall incidents in history which resembled the situations he found himself in personally. "This reminds me of the time..." was heard from Adam a lot. Sometimes it was "I think General Grant would have...," or "Julius Caesar once...," or whomever. History was a very frequent guide for Adam Thorne, in terms of precedence, or of morality. This is a vital factor to consider for those who may wish to understand his make-up. The history of aviation and space travel was perhaps an even more important topic for him. As an aerospace engineer, he was especially interested and aware of everything that had gone before Uplift. This was the kind of trivia he was always able to expand and improve upon. In this way, he could weed out good ideas that had been hidden away for years but never acted upon, whether because they were erroneously thought to be bad ideas, or simply because no one could afford them. This was, to him, just the ongoing pattern of all human progress, especially in the realm of technology. The ideas that ultimately became successful were simply just the convergence of opportunity with a person who decided to seize that opportunity. This was not a bad outlook for someone with the resources Adam had inherited.

As the vacation period drew to a close by the end of August 2085, those who had been away found their way back to Truth or Consequences, NM. Despite his famously secretive nature, Adam sent a companywide e-mail to all Uplift employees to announce that all were welcome to attend a wedding on August 31st at his home. This was issued on the 30th. The time would be 12 noon, and all employees not otherwise engaged in vital activities were welcome to attend. In fact, there would be two marriages performed simultaneously. The first couple would be Adam Thorne and Matilda Orona, and the second, Manjeet Singh and Junaki Sato. Refreshments would be provided, and each employee could be away from their duties until 2 pm if they were working that day. Adam and Mattie had no outside relations who could attend even if they had wanted to, but Manjeet and Juni both had families who were actively employed on site. The local Sikh religious leader would officiate for both couples, having waved any sort of religious prerequisites when he agreed. This ceremony was performed on the lawn in front of the Thorne estate, which Adam had arranged to be lavishly decorated for the occasion. The vows were performed in tandem for the two happy pairs, and both of the final I-dos were spoken at precisely the same time as a Shrike had left the runway and flew directly overhead. It would have been impossible even for Adam to plan things so precisely, but it happened just the same. It seemed like an omen, and perhaps it really was.

Chapter 9
Throttling Up

Compiled from meeting notes and commentary by
Nathaniel Floatingfeather…

Now came September of 2085. Adam convened his "Task Force for Sustained Orbital Operations" on the first Monday of the month. One may notice that actual meeting minutes show that the name of the task force tended to change slightly from meeting to meeting. This is because the group never made a single motion during its entire existence about what the group name actually ought to be. This was an odd oversite for a group of such consummate professionals, but it did prove that they were only human. As with the last meeting, Adam asked for an update on any new thoughts or developments that may have emerged since the last meeting. "Oh, and also," Adam added, "I'd like to introduce Sharon Layton." This young lady was sitting on the side of the table on Adam's right, at the very far end. She was a brunette caucasian who appeared to be in her early 20s. "She'll be responsible for nutrition and food production on our project," Adam announced. Sharon had been recommended to him by Mattie, who had done an online search to fill her position. She had already earned PhDs in agriculture and animal husbandry from New Mexico Tech in Socorro. Everyone nodded to her and said hello.

"OK," Adam said. "If we can just continue with our format from the last meeting, why don't we start with Manjeet."

"Very good," Manjeet acknowledged with his usual stoic seriousness. "I have begun drafting some very preliminary blueprints addressing the habitation ring we had envisioned last time. These have been augmented following the completion of the hotel project three months ago. One thing that has occurred to me is that we have two potential modes of artificial gravity production depending on whether we are simply in Earth orbit or whether we are proceeding in a linear fashion under constant thrust."

"Oh, right, I was thinking about this a few weeks ago, but I can't remember why I thought it was important, let alone what to do about it," Mattie interjected.

"Well, when it comes to day-to-day living spaces, it's very important, "Manjeet resumed. "Let's assume we are in Earth orbit at a constant velocity. I think everyone here knows we would just spin our circular ring at a constant rate in order to push our bodies to the outside of that ring. The outside walls would then act as a floor, and we could walk around the circle in a moderately ordinary way. Agreed? Good. Now, what if we choose to leave orbit and accelerate away from Earth at an ever-increasing speed? In that case, our acceleration would act as our gravity, pushing us against the source of the forward thrust. In other words, in the opposite direction of the forward thrust. If we really want to go somewhere that takes a really long time to get to, we will need to continue increasing thrust for a very sustained period. Slowing down would cause us to hit the ceiling, so we would need to flip over 180 degrees before we hit the brakes. That would continue to push our feet towards the floor with enough force to keep us there. In summary, then, our craft's configuration would need to change appropriately based on our situation at the time."

"Oh my goodness!" Adam exclaimed. "I can't believe I hadn't thought of that! Now that you bring it up, I can't imagine how to deal with changing our internal configuration around when we're already underway! Is zero-G the only option after all?"

"It does get complicated," Manjeet continued. "But how about this? Our habitable spaces are the only really important ones as far as we're all concerned. Water and fuel tank compartments, agricultural and mechanical spaces really don't care about gravity. Even areas like the bridge and temporarily occupied spots like air locks and computer storage areas may not really matter. I'm talking about personal living and communal areas, maybe about one-fifth of our modules. What if they could be rotated along the circumference line within the circle? We could sort of place tubes within the tubes if you see what I mean. They could fit together just tightly enough to allow the inner ones to rotate freely. We could place motors at the joints between the modules and rotate the inner tubes to optimal angles at different milestones of the flight."

Gregor Levinski, the other systems specialist, was always one to see the big picture of schematics. You could see him working all this out in his head as he spoke, almost to himself. "What about the relation to the hallway tube? You can't rotate one without the other, and that would disrupt how you would need to access the modules from the hallway.

"Hadn't gotten that far yet, I'm afraid," Manjeet admitted.

"Well," Gregor mused on. "All I can think of is yet another hallway tube. The necessary modules, let's call them pods, could have two doorways at 90 degrees from each other. One could be sealed when the other was in use. When in orbit, or when acceleration is more neutral, we rotate to the parallel ring and use that door to the first hallway. When accelerating forward, we rotate the pod to a hallway 90 degrees to the rear and switch to the other door. This would necessitate moving the interior fixtures as well, so a wall would become a floor and back again as needed. I'm not sure how that would occur. Some very creative interior rails with wheels or something.

That way, what we usually consider as up and down can be manipulated to their proper places. It should be doable, though. I'll get my engineers on it. That will increase the number of pods by another third. Then again, that's that much more space for conduits and storage. It should help cover all the bases."

"Fascinating," Adam said. "Actually, we probably could get by without two equipment corridors. On the other hand, the extra space would certainly be nice. And it would allow for more internal configuration options. Well, I knew this process wouldn't be getting any easier as we went along. Juni, how goes it on the computer front?"

"Well, now that we have more interior space to work with, it's much easier," she answered. All the equipment can be distributed throughout the various circles of storage space. That will help with heat distribution, at least. Maybe the heat can even be harnessed and distributed throughout the ship. That's a ventilation systems problem, though. We don't need to discuss it today, I don't think. Anyway, if we start now, we can begin collecting data for the huge library you were imagining. I could fit the entire content of YouTube, music, literature, science, whatever we can think of if you agree to pay the royalties. It will take a year or two. We can still upload information into the memory while we are in orbit. After that, it will be more challenging."

"What can we do to expedite it if we leave?" Mattie asked.

"Well, that depends on how far we go," Juni said. "If we're real, real gone, I propose releasing a trail of signal-boosting satellites at regular intervals. It won't really speed the signals back and forth too much, but it should at least keep them clearer. Building these wouldn't cost all that much, but maybe we should equip them with small nuclear batteries to keep them going for a long, long time. It would keep them from freezing in deep space as well..."

"Get some people on that, would you?" Adam suggested. "Anything else?"

"I'm kind of taking this on without permission," Juni finished, "but I'd like to put a little energy into a few quality-of-life concerns. Entertainment, I mean. I don't think anybody wants to spend months in orbit, or anywhere else, without some decent distractions. I think we could spare a pod or three for those kinds of activities."

"Such as?" Peter Hosokawa interjected.

"Well, heck," Juni answered. "Big screen televisions are not too expensive, and they can get pretty big. Remember those "Holodecks" on Star Trek? Why not have a pod or two with flat screens on all sides? We could use the computers to project images that completely surround us. We can't transport solid objects into it like on Next Generation or anything, but we could project images of Earth habitats, movies, whatever. It might help keep us anchored emotionally to our home planet. We could have sound rigged too, for whatever we wanted. Concerts, dances, whatever. Probably not enough room there for everyone at once, but it would be way better than nothing. Also, everyone's private living spaces would have TVs, personal computers, music, and communications capabilities. Assuming we authorize it and get the funding.".

"Now that you mention it, I don't see how we could do without it," Adam said. "You know what I was wondering?" he asked. "Does anyone here play any musical instruments?" It turned out that Iza Mazurkiewicz played some violin, and Azaan Nazari played drums when he was younger. "I think we should carry an assortment of instruments with us and instructional materials, videos and the like, just in case we get really bored."

"Maybe our kids could learn as well!" Sharon Layton shouted enthusiastically. This was her first contribution to the conversation, and it caused a sudden unexpected hush in the conference room.

"Well, we haven't dared to breech that subject so far in this group," Adam said rather guardedly. "Obviously, we will have to address it eventually as we progress. But for now, I think we should just keep trying our best to come up with a space vehicle, or station, or what have you, that will keep a moderate crew of people alive for a while. Keep reminding us, though!" Adam said. "Well, Peter! Oh, excuse me, Dr. Hosokawa, could you give us a report?"

"Decided to call me by my chosen title, have you?" Peter responded rather flatly.

"Yes, I have," Adam countered. A small amount of clapping was heard from several in the room. "Mattie has mentioned to me a time or two since we last met, that I was out of line last meeting by demanding you be called Peter. I've decided to recant. Now, I'm a bit of a history buff, as you may know. I often decide to use historical precedent as a guiding theme for this little group of ours. It seems to me that what we are considering doing resembles an early sea voyage from the days of the first explorers. Those ships set out into oceans that had barely even been charted, not even knowing whether they would fall off the edge of the world. Just imagine how terrifying that would be! The only thing that kept them on mission was a system of conduct and mutual regard. Now I'm not saying I'm appointing myself Admiral or anything. I'm just acknowledging that you were correct about respecting each other's dignity and abilities. Actually, I think, assuming you all want to participate in whatever the mission we decide to undertake, that we should have a big heart to heart once things are all decided and agree on a code of conduct for everyone to follow. Now, Doctor, this would be intended to ensure that no one would be able to assume some sort of dictatorial control over the entire group. We wouldn't be granting anyone command authority without a general agreement to do so. There would still be specialized subgroups of crew responsibilities, including our medical concerns. Those involved in particular specialties should be able to choose their own leaders as well. But in the old days, ship's Doctors had rank, and were called "Doctor" out of courtesy, and to maintain at least some sort of social structuring that everyone could recognize. I think that should go for everyone. If Mangeet wishes to be referred to as "Mister Singh," that's what he shall be called. As for me, Adam is all that is necessary. We should all choose what we wish to be called. Sorry to pontificate, but I need you to know my apology is sincere. Although, Dr. Hosokawa, you should be aware that tension between flyers and Flight Surgeons goes back for decades. Especially true for astronauts."

"Yes, that is rather well known," Dr. Hosokawa chuckled. He appeared a bit flushed. "But thank you. I'm not used to anyone apologizing. But I will go ahead and request the use of my title from all of you, at least until I say differently. That being said, I do have a rather lengthy report."

"Now's your moment," Adam laughed.

"Very Well," Dr. Hosokawa commenced. "Just remember you asked for my opinion. To start with, I've polled several of my former colleagues from my last orbital project to see if they might want to join this one. Not all of them actually ever got any flight time, but they all said that they would love to try and get some. My first choice was a general practitioner named Cynthia McKesson. She's a fairly young new Doc who could easily manage the pharmaceutical side of things. When I told her the focus of our project, she rolled her eyes and asked how much money she could have. She also said her husband Larry actually is a pharmacist, and he says that he would love to be part of any sort of Space related project. Anyway, I went ahead and described a theoretical voyage lasting decades in duration, with a crew of say, 20 souls. She thought about it, then said that would require a supply of medications comparable to what a large metropolitan hospital would need for at least a month. Now think about it, imagine if any of us young adults developed diabetes at an early age or arthritis or asthma or hypothyroidism, or any number of chronic diseases. What about allergies, heart disease, or any number of other problems? Those people would need a lifetime supply of certain medications. Should we just tell them that they're out of luck? That would be obviously immoral, not to mention detrimental to the mission itself. Anyway, neither of these two seemed to seriously consider joining our task force."

"Apparently not," Adam said. "Their suggestions were relevant, though. So, in addition to our baseline medication supplies, could we outfit ourselves to manufacture more if our supplies got low?"

"Absolutely, with enough money. We'd also need space for both a pharmacy and for the necessary equipment."

"Submit a purchase order to Frieda, my secretary," Adam said flatly. "For equipment and all recommended necessary medications."

"Whoa, hold on!" Mattie interrupted. "You make it sound like we're leaving next week."

"True! But we might as well act like we are, just to get things moving," Hosokawa agreed. "Maybe we could start with the equipment, but we still haven't compiled a list of best medications. We haven't talked about chemotherapy drugs, surgical anesthesia and equipment, plasma, vaccines, or any of that, but I guess I just did. Now, one thing in our favor is that we probably won't be leaving while carrying any active infections along with us, and we won't be coming into contact with any once we leave, if we ever do leave. Anyway, we'll still be needing stockpiles of things like shingles vaccines, Tetanus vaccines, and vaccines against anything else we might be carrying latently, you know? We also need to be cautious if we ever run across some planet or moon that we'd like to inhabit. If it already has lifeforms present, we'll be bringing our own set of skin microbes and intestinal flora which will therefore infect the planet. Remember how the Apollo Moon mission crews had to isolate themselves for observation. We might need to isolate ourselves to protect the new habitat. No one ever seems to think about that."

"Well! You are thinking ahead, Doctor," Gregor, the systems specialist, blurted. "We'll need to coordinate on that. I wonder if we'll ever need to implement any protocols in that regard."

"I have no idea," Hosokawa answered. "We just need to keep very open minds just now. Anyway, let's move on to OBGYN concerns. As it happens, I've become quite involved with a physician named Xiang Xao Jia. She's from China and appears quite intrigued with the problems I've described to her regarding this project."

"Involved?" Mattie asked. "How do you mean?"

"Well, romantically involved, now that you mention it," the Doctor answered.

"Is she likely to want to join the crew?" Adam couldn't help but ask.

"Well, again, if she does, I will as well. It's way too early to really think about any of this."

"Fair enough, go on."

"Well, as I had mentioned last time, a birthing experience in low gravity certainly poses more problems than we can probably anticipate. The more gravity, the better, in terms of capture of the fluids released if nothing else. We will need ICU level equipment for the mother and infant as well, even if the baby is premature. If we are still planning on a pure oxygen environment, that might be beneficial at times like this, otherwise, supplemental oxygen should be available. That brings us to a need for a fully capable surgical suite. We ought to be equipped for everything from an appendectomy to heart and blood vessel catheter placement to cancer interdiction. I am personally certified for all of those types of surgery. But again, we'll need a full radiologic capability. That means an x-ray facility, and fluoroscopes for cathing procedures. An MRI machine wouldn't be a bad idea either. These things are not quite as heavy as they had been in the past, but they're still pricey. I think if they can be tied to the on-board computer suite, at least we won't have to recruit a radiologist."

"Yes, I've been planning to upload a full medical database that is meant to be tied in!" Juni was quick to respond.

"Excellent," Hosokawa said. "Now, I should alert Mary here that in terms of on-board manufacturing, I suppose we can never tell if we might need things like artificial heart valves, or artificial joints for replacements, or even surgical rods for mending fractured bones, that sort of thing. An orthopedic surgeon aboard would be great, but more crew means more mouths to feed and all that. I can tell you that computer-based advice on best practice procedures would probably be adequate to prepare me to perform that sort of surgery if it ever becomes necessary, as long as I have the resources. I think the same would go for ophthalmic and dental surgery. That brings me to equipment for eye care and dental care. Xiang could be my assistant for most things, but I wonder if an on-board dental and maxillofacial specialist wouldn't be a bad idea. Mary, you might even be called on to produce things like eyeglasses and dentures for us."

"Well, yeah," Mary Dupree joined the conversation. "While I'm talking, I can go ahead and add that I would like to request two large scale 3D printers that can manufacture very large replacement parts for the station. These don't require a gravity environment and can be placed outside of the habitation ring if necessary. All that's needed is easy access for the crew. I'll be asking for one or two smaller units within the ring to manufacture everyday items. I guess the corollary is that I'm planning to join the crew as well."

"This makes me wonder," Adam asked. "If any of these theoretical on-board children ever need braces, can we help them with that?"

Hosokawa chuckled. "I think we could be of service, yes. I suppose we'd go with the clear plastic mouthpiece type that they're using these days. No metal hardware glued to the teeth that you still see sometimes. Parents would still have to make sure they are being used properly, of course. But we should still be able to do caps and tooth implants, all that kind of thing. At least nobody's going to need health and dental insurance to stay on board our little ship. Anyhow, I hope this gives everyone an idea of the complexities we're facing. Unless you'd all prefer that I just play the part of that old sawbones you talked about. I could just do amputations and pull the occasional tooth now and then."

"No, Doctor," Adam reassured him. "This is an era where all these concerns are completely relevant, vital even. No one would care to spend a large chunk of their lives confined in an enclosed space without a good chance to live in good health and comfort. That's an actual fact. Anyway, what's the point of traveling off into the unknown, theoretically of course, if no one survives the trip? I'd say push on, full speed ahead with your medical recommendations, and keep us and the finance department up to date on what you need. A good starting place would be how much space on the station you'll think you need to contain your on-board department. Oh, and I think we'd like to meet your new, um, associate. Maybe at next month's meeting. Ok, Gregor, can you speak next?"

"Da" he answered, forgetting his English. "Oh, yes, I'm sorry. I really have very little to contribute just now. This idea of internally rotating our cylindrical pods to meet the forces of acceleration is quite insightful. I will work with Manjeet on this, yes? Good. Now something I've been thinking much about is this idea of maintaining a pure oxygen environment. I believe it's historically been an expectation to do this merely because anything else is considered too complicated. True, if we have a very large water supply on board, then oxygen can be readily available for a very long time. But I think a large supply of liquid nitrogen can also be brought on board. It can be stored outside the main wheel in liquid form and heated to a gaseous state when brought inside. I think adequate regulation of the gas balances in the habitable spaces can be achieved with current technology. Most of what we're breathing now on the ground is nitrogen, after all. Nitrogen isn't metabolized or wasted, and therefore would not be considered a consumable. The supply would be constant unless there were some sort of leak. We could maintain air pressure at what we are used to here on Earth, an oxygen-nitrogen mix, and the risk of fire would be lessened. Not that we wouldn't still need to maintain high vigilance for fire safety. That's always the highest priority for any ship, even on Earth. Now, Mangeet, I think we should spend every free moment we have on designing one of these habitation modules, or pods, did you say, as soon as possible. Then, we can get with Mary and see if we can actually build one. What do you say, everyone?"

"Absolutely, do it!" Adam said. "Tell us how you're progressing one month from now. And don't be surprised if I pop in on you every so often. A pod would definitely be a good first step into reality. Thanks to you both. Now, I think we're all anxious to hear from Sharon."

"Thanks, everyone," Sharon commenced. "I can't thank all of you enough for giving me the opportunity to be here with you today. This is a fascinating field of study, but the furthest I got back at NASA was to design a few pallets of seedlings to send aloft to monitor for a few months. Not that that's bad but considering how to feed a small group of people in a spacecraft ad infinitum is an entirely new field of research. Now, I suppose I'll state the obvious to start, just to get that out of the way. From what I'm hearing, there should be a pretty large amount of space that we can devote to storage of foodstuffs and keeping these at low temperatures would of course be the best way to preserve them. Refrigeration would obviously be easy to maintain in space," she giggled to herself. "Therefore, I suppose it wouldn't hurt to stockpile some items that it would be hard or impossible to raise aboard. That would be things like beef products, poultry, and pork. I've heard that Adam had at one time considered orbiting some cattle in the past but decided against it. Good choice, Adam!"

"Amen!" Mattie interjected.

"But he did conclude that orbiting fish was not a bad idea. I must say I concur with that plan wholeheartedly. Plenty of human societies have been, and continue to be, sustained with both fresh and seawater food sources as their primary source of calories and protein. I imagine that our water storage pods can be adapted to both fresh and seawater containment?" Gregor nodded in the affirmative. "And at a variety of temperatures?" Another positive nod. "Excellent! There is really no need to limit the number of species, vertebrate or invertebrate.

Shrimp, octopus, snails, cod, salmon, lobsters, whatever. Probably, the more variation, the better. A little plankton? Why not. We can put that at the bottom of the food chain so eventually everyone can all feed themselves, and then we can eat them. I can certainly train whichever crew members are willing to learn regarding how to maintain the tanks properly, as well as maintaining the ecological balance. The more interconnecting pods we have for them, the more they can swim around between them and procreate. We can have sand and rocks in place for habitation, well, at least as long as there is some sort of artificial gravity. I assume we can rotate those pods the same way we rotate our own? Well, I'd recommend that if possible. Now finally, I think you all have heard that it's possible to sort of clone animal tissue in tanks to produce edible food items these days? Well, it is. If we carried the appropriate tanks and equipment, we could do that ourselves in isolation. It would be kind of a last resort just in case we find ourselves getting low on anything. We could use what leftovers we have and generate more edible material. That would take another pod, of course. I could instruct crew on those procedures as well."

"Now," she continued. "We can move on to the vegetable components of our diets. Obviously, humans have been growing all manner of plants in zero-G for decades. Now it appears we will have room for far greater production than has ever been accomplished. If we have large stretches of corridor and storeroom space, I propose we should use it whenever the opportunity presents itself. Plants generally do not complain about a lack of gravity, it appears. Useable light and water, as well as nutritious soil, are all that is needed to grow nearly anything. This has been proven time and again. Therefore, we'll need to launch a good deal of soil into space. If this isn't deemed prudent, then we could go hydroponic, but it hasn't been seen to be a problem on other orbiting platforms. I don't think we should limit ourselves in any way as to crop variety, except maybe in terms of available storage space. No giant Redwoods aboard, please, unless we bring seeds to plant on some remote world. Plants for our aquatic companions obviously should be provided as well. In fact, a tank or two with lots of Algae and seaweeds to dispense is a must. Now, I probably don't have to reiterate that plants do great service in an enclosed environment. The more plants we have aboard, the more CO_2 they can consume to make their own glucose and everything else, and the more O_2 they can emit as a biproduct. This can help transform our ship from a mere machine to a true ecosystem in its own right. There's your sustainability right there, Adam!" I assume that a good deal of your crew's time and energy will be spent farming and fishing, as it were, in addition to being mechanics and pilots."

"And just to show I've really been thinking a lot about this," she summarized, "I'm sure it's not an unknown topic to you, but if you, or maybe even we decide to take this beast we're building away from this particular planet, all of these things can be exported with us. Now, nobody has the faintest idea of what we might find out there. If it turns out that there is some barren rock that has its own water supply and atmosphere, as well as a survivable temperature range, then we can bring our little biosphere down with us. We should have more than enough seeds and aquatic life forms to spread around to keep ourselves going in perpetuity, whether we stay or whether we skedaddle later. On the other hand, if there is already life established there, we can at least investigate to see if whatever we bring is compatible with what we discover. That goes for Redwood seeds or our intestinal flora. Don't you agree, Doctor Hosokawa?"

"Yes indeed!" he replied. "But I must admit that accomplishing any of this seems the most far-fetched possibility I can imagine. Even actually building a craft of this complexity is surely a preposterous proposition. But then again, Columbus himself was seen as preposterous by his own crew, if I'm not mistaken."

"Nevertheless," Adam concluded. "What Sharon says rings true to me. Let's see if you can write up some quantifiable proposals for us for next month. Gregor? Could you spend some time with Sharon and see what might be required in terms of heaters, pumps, water purification, and so forth for her aquarium pods. They also have to be accessible from the service corridor for when we need to go fishing or get to whatever we need to eat. You know what I mean."

"We're already thinking about it, boss," Gregor said.

"Ok!" Adam said. "I think then that we'll hear from Azaan, if that's OK, then maybe we'll adjourn until next month. Is that alright with you Azaan?"

"Good a time as any," was the response. "This is all pretty thorny, but we'll have to deal with it eventually."

"Yeah, I'm sure we've all got some of this on our minds. Go ahead," Adam nodded as he spoke.

"First of all, I'm feeling that we shouldn't limit ourselves to one mode of propulsion. Hopefully a cumulative approach might shorten our transit time a bit. We've heard this already. Plus, it won't be a bad idea to have backups if something goes wrong with one or more of our engines. I'll start with a proposal to build, I think, two ion drive engines. I mentioned this before. These use a strong magnetic field to shoot out ionized gas from the rearward facing end of a chamber at mega high speed. I think I said argon gas at our last meeting, I can't recall. Most literature actually recommends Xenon gas, mostly due to its suitability to be stored in liquid form at the temperature ranges present in deep space and so forth. But Xenon is really not widely available in large quantities and is also extremely expensive. I think we could probably use something cheaper like Argon and just rig the storage tanks for whatever temperatures and pressures we choose. Now a lot of electrical power would be needed to drive these beasts. NASA has used small ones to send out probes within the solar system, but no one's ever used anything as big as I'm imagining before. The NASA probes tend to use solar panels for electricity, but I think we would need to use some fair-sized nuclear reactors to generate the wattages required. Not only that, but we would also need one to power the internal equipment of our spacecraft for the time spans we are envisioning. But more about that later. The thing about ion drives is that they build up their thrust at a very slow rate. You don't just turn them on and start at a sprint. I think we'd need to leave them on constantly for about two years to get up to some sort of maximum velocity, but at least it would build up at a constant rate. Then we'd need to turn the whole ship once we get to maximum velocity around and fire in the opposite direction to slow back down. Now, while we're close enough to our sun, we could also use solar panels to generate electricity. But they would be useless for much of the trip, at least until we got close to another star.

That would also be true if we used solar sails to produce forward momentum. Those would be need to be pretty huge, but cheap as well. No matter what our course, solar winds radiate pretty evenly through the entire spherical area around the sun and could be used to propel our craft in any direction we choose away from the center of the sphere. That would also build up our momentum on a slow but constant basis, and we wouldn't slow down until we got to our next star, which would then start slowing us down. Solar sails are obviously meant to augment our propulsion; not be the primary source. Now, the other major option, which will assist and or replace the ion drive, would be an atomic engine. This would use a nuclear reactor placed on board to massively heat a flow of gaseous fuel, hydrogen is the most recommended, and use this to create superheated plasma that would shoot out through a nozzle to create thrust. This hydrogen could be stored as liquid and superheated as it approaches the core. Of course, the immense heat generated by this process is the major problem with this method. If it gets out of control, then we could wind up with an atomic explosion that destroys us all. Computer models show that internal structures using infused boron can assist with the control of heat and more closely control the fission. Good old-fashioned control rods can assist as well. The fact that we are in deep space also can assist with heat dissipation. What we will need is a healthy amount of shielding between the engine and everything else. The central frame of our good ship should be very strong, and rather long as well. The engines should be way at the back, with large fuel containers and pumps clustered behind the engine pods. The fuel for the ion engines would also be in temperature-controlled tanks placed wherever they will fit. Now, somewhat conveniently, we could use water electrolysis to generate hydrogen and oxygen. But I think entirely separate supplies of these gases should be stored in the rear for fuel purposes only. We'll need to start structural analysis tests to see how best to arrange all this around our stern. We already have plans regarding how to change which way we're pointed and so forth. Anyway, this is all a major project. Scientists have been theorizing about all of it for decades. No one has tried to execute any of these options yet because of the major funding it would require. In addition, no one has found any reason to try it since there probably wouldn't be any reward for doing so. Who wants to shoot themselves out into deep space without a definite goal to justify the journey? Nobody so far. I don't know that that's not what we're trying to do now, actually. At least our manufacturing methods are much cheaper lately, so we won't waste as much money. It's just that really the only way to test it all is to just go up and try it."

"Quite a lot to think about, that's true," Adam agreed. I'm prepared to go forward if you are. If your team can draw up some blueprints that computer models endorse, perhaps we can start manufacturing some prototypes for these things. It will be a start. I'm talking about each and every option you've just mentioned. Please remember that this is all proprietary technology. The patents will go to the developers, of course, meaning yourselves. I can only say that I hope you don't take your designs elsewhere, since Uplift is doing the funding. If we do take it all off-planet, it won't matter much, of course." Audio recordings indicate that at least a few of those present actually giggled at this. "Now, the radioactive elephant in the room deserves a bit of discussion. It now seems that regardless of where we turn, nuclear power is the solution to every problem. As we also know, national policies and a bevy of treaties currently prohibit sending nuclear materials into space unless there is little to no orbiting of said materials around the planet. If you just shoot it straight to wherever you want to go without any significant time twiddling your thumbs over the local inhabitants, it's not too much of a problem. For our project though, it appears that we will have to use conventional power like solar panels here around the Earth. But if we want to employ the other power sources we've talked about for any length of time, we'll just have to leave the neighborhood. The other major problem would be how to realistically acquire that much nuclear fuel for our own personal use. I'd presume that something like weapons grade Uranium or Plutonium would best suit our needs. But I can't begin to imagine how we could get our hands on any of it, even using nefarious means. Granted, if we did decide to steal it, we'd have the ultimate getaway vehicle. Oh man! I can't believe I'm even saying this. Maybe we could ask for permission from the government, but would we get it? And what would any government have to gain by funding us? It would inevitably cost more if it were a government project anyway. It would be way cheaper for a private company that already has the infrastructure and financing available, like us, to do it. That way, when the money's spent, we'll just send it into deep space and leave them holding the bag? Hmm. This could really get complicated. Really, for the time being, I think we should just keep all of this between ourselves. Otherwise, somebody's definitely going to tell us to cut it out. We'll have to watch our backs, that's for sure. Anybody else have any opinions?"

"Yeah, we all do," Mattie replied. "But nobody wants to stay in this meeting all week. On the other hand, I do wonder what to tell the world when we start building this heap of orbital junk and have to say why. We can't just tell the planet to mind its own business."

"I guess we'll just have to call it a research facility," newcomer Sharon volunteered. "They wouldn't believe anything else anyway. If we decide to head out, we just won't tell them. There won't be much they could do about it anyway, even if they wanted to."

"I'm afraid to say it, and no offense to the rest of you, but that may well be the most reasonable thing anybody has said here today," Adam admitted. "And that's saying a lot. Pretty scary to think about though. Thanks to all of your for thinking bravely. Well, all right then. Keep on planning. We'll meet again next month. If anyone feels ready to build anything, get with Mary and I and we'll see what we can come up with. Thanks, all."

Chapter 10
Audacity? Sounds Fine!

Compiled from recorded Uplift meeting records and US Government records...

In late September of 2085 then, Adam requested a meeting with the Board of Directors at the Federal Aeronautics Administration (FAA) in Washington, DC. FAA inspectors were already a nearly permanent presence at the Uplift corporate headquarters after Frances' accident. These government employees had the run of the facility, checking that personnel and equipment were up to government standards. Inspections were usually done on a random basis and also without even the slightest warning. These Feds even had a small field office located in one of the hangers on the flight line. In all truth, Uplift really did have an impeccable safety record, the glaring exception being the incident with the influential Frances. Therefore, these FAA officers were generally regarded as people with very little to do. After a few years at the site, they were really just considered a necessary facet of the daily routine of the company. Most of the officers would soon find themselves on a first-name basis with just about everyone at the compound. Conditions did change a bit with the hotel construction project; however, as an entity from outside, Uplift had commissioned and supervised the work. Endless Horizons, Inc. also had its own legal department to deal with compliance with FAA regulations and statutes. In general, any nation or company wishing to use Uplift to place objects in Earth orbit was welcome to do so by the US government and were not subject to much regulation once they were in orbit. On the other hand, every nation that engaged in launching such cargo had their own rules regarding what was launched from their own soil. Some items were actually strictly regulated. Weapons were the major concern. Nothing that could be considered potentially harmful to other nations was allowed, including explosive or chemical agents, and especially, of course, nuclear weapons. This was generally the language to be found in most US treaties, but obviously, if one nation was at war with another, there was no real way to enforce these treaties. As has been noted, this would, in practical terms, apply only to items in payloads that would orbit over other nations. Payloads departing directly away from Earth into deep space or theoretically directly placed into geosynchronous orbit over their own parent nations of origin were allowed. But that would not be a problem since no nation would wish to have harmful payloads orbiting directly over their own citizens.

When Adam realized that the huge space platform he wanted to place in orbit in the near future would undoubtedly attract the attention of the US government, he decided to act preemptively and basically warn the FAA agents in advance. As Sharon Layton had suggested, he decided he would call the station a "Research Facility," which in fact it would be. He was well aware that, as he had mentioned in so many words at the August Task Force meeting, he would never in a millennium be granted permission to place large amounts of nuclear fuel in Earth orbit, even for a few seconds. At the Washington meeting, he outlined everything else he planned to do, even to the extent that he advised the directors that he might well place prototype engines on the station "simply to see how the hardware withstands the space environment." None of the directors actually showed any suspicion that he might not be being completely transparent on that topic, but they did reiterate the standing regulations just described. Overall, since no government money was being requested for the project, the government really had nothing to say about going forward with the project. They did warn Adam to expect that the on-site FAA inspectors would begin to make their presence more obvious, and there would certainly be more of them. They also stated outright that the agency would be extremely interested in any technology that might benefit the United States that might be developed during the project. Finally, they would be on the alert for any spying or espionage regarding the project that might develop under Uplift's radar. "You just never know who might be skulking around," they said at the meeting.

On returning home, life continued with apparent normality at Uplift Aeronautics. Adam and Mattie later remembered this period as especially pleasant. Though Adam's new plans and goals always kind of boiled in the backs of their minds, there was at least time for privacy and private interests on the weekends. This was apparently the only time in their lives that Mattie and Adam actually travelled a bit around planet Earth together. The last two weeks in September they even flew around Europe, quite aware that might be the only time they could see anything there at ground level, and not from orbit or the internet. Manjeet and Juni also recalled that time period as especially pleasant. By the time October arrived, however, Uplift business returned, and Adam was invited to the Uplift production plant near Grants for a status check on the habitation pods that Manjeet and Gregor had been designing. Actually, the whole task force wound up attending since everyone wanted to give their input. They flew over from the launch facility by helicopter. It seemed quite remarkable to everyone that a prototype had been built in just a few weeks, but the use of computer design and modern 3D manufacturing processes were making such feats rather commonplace by that time. For example, whole slices of an oceangoing cruise ship could be pieced together in a few weeks, not months, and then be assembled into a completed vessel in fairly short order. This pod they were inspecting was really just a football field sized sausage made of composite materials. They were manufactured with included ducts, conduits, pipes, wires, and a multitude of other surfaces where all manner of supplemental parts might later be attached. It was basically just printed out, per computer specifications, in a "collapsed" state, then pressurized to expand to its final cylindrical shape. This new demonstration pod had already been filled with the appropriate gases to a pressure of one atmosphere, then connected to an air lock that they could enter through single file to begin their tour. The whole group was escorted inside in small groups to closely inspect the pod. Manjeet and Gregor were justifiably proud of what they saw. The pod had been filled with wooden mock-ups of various furniture items that had been suggested, like a dining table, a kitchen set, couches, a television screen, and one twin-sized bed. The rail hardware that would allow the entire interior to spin in either direction inside the outer shell was simulated but was not yet in its final form. But since the pod was currently on the surface of the Earth, the actual spinning could not be demonstrated properly yet. Manjeet said he felt that every crew member, or perhaps a "family unit," could be given two adjoining pods. One would include a kitchen with food storage capability, as well as a living room sort of space. The other would have beds (or hammocks if G-forces were insufficient at that time in the flight), as well as a "bathroom" for the pod's occupants. Gregor stated that a suction-based toilet similar to those used on previous orbital stations would be located in these rooms. He even felt he could design a shower stall that would dispense warm water from the top and suck it downwards into a drain system at the bottom. This most likely could work even in zero-G, although he hoped the showers would not be too prolonged. A water treatment tank would be available in one or two of the crew-ring pods to filter and purify the water for reuse. Urine would also of course be reclaimed. Solid waste would be recycled as well, as this would be Sharon's best source for plant fertilizer. Gregor was quite enthusiastic about all of this. A variety of "dressers," or at least lockers for personal gear, would be standard in each bedroom pod. Gregor stressed that by bringing a 3D printer aboard, crew members could even express their own preferences by designing their own furnishings if they desired.

Adam gave his blessing to everything he had seen on this inspection stop. No one else had anything negative to say either. He suggested that the manufacturing crew begin producing the pods in bulk just as soon as they felt ready to do so. There would be a brief period while supplies were ordered and then delivered, but this would only entail perhaps a few weeks before construction could begin. The interiors would be designed individually for each crew member as soon as the occupants filled out an online questionnaire. But all of that would be installed after the pods were aloft and connected. Furnishings would be brought in via air locks through the interior corridors. Everything would be bundled up in packs that would fit through the hallways and hatches and then assembled inside the pods. Later that week, Adam met privately with Azaan Nazari again regarding propulsion options. He felt that the fewer people who knew about any specific plans that were agreed upon, the less chance there would be for unwanted leaks. Such whispers might find the ears of the new FAA agents at the main flight facility, thus jeopardizing any chance of developing anything. They actually met at Adam's home in the guise of a social visit, but Adam always took notes, which he put in the task force archives. He even did this at actual parties. Azaan reported that he had already put in an order at manufacturing for a set of two quite huge solar sails. As the means and place for mounting these to the station hadn't yet been finalized, various mounting devices were designed that might be used when the time arrived. That would all be decided later. He also took it upon himself to order some large solar arrays for power generation in places where a nearby star was available. Smaller units were ordered and were meant to be directly mounted on the outside of habitation pods as well. Adam would be getting the bills, which actually weren't that large given the wide availability of solar panels, as well as for the solar sail Azaan had designed. Next came the harder parts.

Azaan informed Adam that the chosen onboard reactors used to generate electricity for the habitable areas of the station would be the same type used for Naval vessels throughout the world. The US design was generally felt to be the most reliable, never having caused serious safety breeches or casualties despite many decades of use. The problem was that the specifications for these reactors were highly classified and not generally available to the public. He confided, however, that he had managed to get some blueprints for them from the Tanzanian embassy. Tanzania had actually purchased a few nuclear naval vessels over the last ten years. These submarines were based in Zanzibar but were unfortunately not accessible for inspection. Azaan had joint US/Tanzanian citizenship and had contacted the Consul in Los Angeles discreetly to make inquiries about these reactors about two months previously. He had mentioned that these inquiries were related to research that Uplift was considering. He had just recently heard back from the Tanzanian ambassador. Apparently, since the technology had been purchased from the US Navy in the first place, there wasn't really any objection from the Tanzanian government per se. This was rather dubious logic, but Azaan decided not to split hairs. The embassy had gone ahead and forwarded an accompanying request made to "Enable an agreement to proceed" with the inquiry. In return, Tanzania was requesting to buy two Shrikes from Uplift, along with support equipment, for a rather low price of 420 million dollars. With this agreement, nothing regarding the reactor blueprints would be disclosed to the US government or anyone else, but they would ultimately be provided secretly directly to Azaan. The actually nuclear material needed to power the reactors was none of Tanzania's concern, the agreement said. All of this was obviously rather disquieting to Adam, although he undoubtedly realized far in advance that some cloak and dagger type deals would be necessary for his more unorthodox ideas to come to fruition. He later said that he "Sighed heavily," when he heard about the scheme, but gave his OK nonetheless. The details took a couple of months to work out. But Uplift had been exporting technology to US allies for many years, and in general the USA had no problems with increasing ties to friendly nations. Since Tanzania was currently a stable democracy, there were no objections forthcoming from the Feds.

Another by-product of this conversation was the decision to establish a new and very secret engine research facility at the Grants production plant. Adam had a new "Storage Building" constructed some distance from the main plant that had a few offices near the main entrance. He had a cleverly hidden entrance installed at the back that was actually an access point to a hidden basement where research and production could be conducted. Crates of mundane production materials were piled up conspicuously on the main floor. Building all of this without government detection was a bit tricky, but it was never discovered until most of its mission was fulfilled. This marked the beginning of Adam's willingness to slacken his observance of laws and rules when it happened to suit his purpose. This was a new attitude for him, and those of us who knew him well would never expect that he could have taken this sort of path. But this was only the first of many devious decisions he would eventually make to pursue his secretive goals in the coming months. Generally, such actions were financed using his own personal wealth and were certainly never disclosed to Uplift management. It remains quite astonishing that no one who truly trusted Adam ever discovered what was going on. They would have been devastated if they had ever found out and may have tried to shut him down if they did. Equally astonishingly, he was also quietly gathering a small cadre of people who had become more loyal to him than to the law. Even now in retrospect, there is still widespread speculation that Adam was actually forming what could only be described as a cult during this period. In his case, however, none of these potential concerns would wind up mattering much in the end. But that's just this author's opinion. None of this would be discovered until much later in the timeline.

Azaan also revealed that he was making progress with his plans for the ion drive engine. He felt that it really would be feasible to use Argon gas for this engine instead of Xenon. Argon actually makes up about 1% of the Earth's atmosphere and theoretically could be stored at the proper temperatures and pressures to make the engine work, although possibly not as efficiently as Xenon. This could be alleviated somewhat by the ability to carry a larger supply of fuel. Adam apparently consented to his proceeding with development. Azaan's greatest challenge of all, however, would be the big atomic engines. He certainly had no Earthly idea how a large amount of weapons-grade Uranium or Plutonium could be secured and brought aboard an orbiting space lab without getting arrested, or even killed, by one government or another. All Adam could say is that once the secret storage facility was built, and it wouldn't take too long, he should go ahead with the overall design and building of the actual engine components. Azaan immediately agreed to do so. Perhaps this was simply because of his professional pride as an engineer, but the ethical conflicts must have churned around in the back of his mind as he commenced. In fact, he did have yet another option up his sleeve, but at that time it was so seemingly fantastical that he didn't dare even bring it up. That would take another few week's research to be that brave.

As time passed during this period, Uplift kept on plugging away with the business of orbital space transport. Company records show that all of those daily missions used to collect and disperse space debris had actually begun to make an appreciable impact on the orbital environment. In opposition to this was the great increase in new orbital satellites that were regularly being launched by nations all around the globe. Paradoxically, much of the new satellite traffic was likewise being launched by Uplift. At least the new hardware now was always equipped with small rocket engines that would be used at the end of any new satellite's useful life. These newer models would then be turned towards the sun and propelled away to be burned up. They could also be kept out of the traffic lanes most often used between the Earth, the Moon, and Mars. Overall, though, NORAD was as busy as ever keeping track of it all. The Sustainable Orbital Facility Task Force, as it was called at that point, had two more monthly meetings where some progress had been made. By the third month, the clandestine engine research facility had been constructed. A bit of work had just begun on Azaan's ideas. About twenty habitation pods had been printed, and interlocking docking ports were designed to connect them that were angled just enough to result in a perfectly circular structure when all the pods were hooked together. These docking structures were to start production full bore in the coming weeks. The single most significant development to occur so far in the project, happily, was announced with this meeting.

Adam was visibly upbeat and energetic at this January 2086 group huddle. He commenced it at 8 a.m., having asked everyone to arrive an hour earlier than usual. After a few donuts and coffee, he made an unusually cheerful announcement to start things off.

"Good morning, everyone!" he proclaimed. "Knowing you all, you may have already heard rumors about this. It was on CNN about 6 a.m. But, to make a long story short, the Endless Horizons, um, motel we constructed last year has already been put up for sale. The company says it just hadn't counted on the massive upkeep costs. It also got a lot of bad reviews due to the high rate of airsickness, vertigo, and so forth. Plus, there were several actual injuries from people hitting their heads on bulkheads and their insurance fees are now greater than any income they've been able to generate from incoming reservations. Apparently mega wealthy people still prefer their yachts as opposed to extended stays in Earth orbit. Not to mention that Uplift has started asking for service charges to transport their guests to the facility. There's been no drop-off in travel companies booking those short suborbital hops though. Just the orbital hotels are losing money. Same as it ever was, I guess."

"Oh, yeah," Mattie added. Plus, we own the company that sells the flight insurance too. Endless Horizons has only explored a universe of endless financial losses." There were just a few giggles in the room with this statement. The group apparently did take at least a tiny bit of interest in matters of human finance. On the main, though, these engineers tried to rise above such trivial human concerns.

"Anyway", Adam continued. "They offered to sell it back to us for a measly two billion dollars and then declared bankruptcy. Obviously, I've gone ahead and accepted the offer as of 9 pm last night." The reaction to this news was some general puzzlement but was generally favorable. Gregor seemed especially pleased.

"So now we have a base of operations for everything else we have planned!" He exclaimed with genuine exuberance. "We have habitable space, some basic infrastructure, electricity, everything!"

"My favorite part is the docking bay," Adam chimed. "Now we won't have to build another. I think that could move us forward about six months with our station project at the starting gate. Let's set our next goal as getting a framework structure aloft as soon as possible and build from there! The hotel will become part of the new overall structure! Mary, Gregor, Manjeet? Can you get with Suni and come up with a general station design? Obviously, all construction materials will need to fit in packages that can fit in the Shrikes and then be extendable. Of course, you all knew that already."

"We sure did, Adam," Manjeet said. "Now, I believe you'll want the new frame to fasten onto the existing structure that's there and allow it to continue to spin. Do you want this frame to spin along with it? Or should it be independent?"

"I'm sure it would improve the flexibility of the design if it were independent," Adam said. "Don't you think? We'll need to provide access through the frame between the hotel structure and whatever we place behind it, namely our much larger habitation wheel. The new large wheel should also be free to rotate at its own rate. The frame in between can be used to anchor all the tanks, antennae, conduits, and whatever else is needed. We can always add things as we go. Everything behind the second wheel will need a continuing framework as well. This will mainly be fuel and engine hardware. But maybe we can also include some large zero-G spaces for things like open space and recreation. The main thing is that there needs to be complete freedom of movement for us to get from place to place inside the station. It's definitely not possible to provide uniform gravity conditions everywhere. In terms of design though, the sky's literally the limit," he giggled lamely.

Iza Mazurkiewicz was there. She was always present for these meetings but seldom was asked to provide any input. For almost the first time, she seemed quite energized by today's revelations. "Did we get the robots with this deal?

Adam responded enthusiastically. "We get four of them," he said. "They kept two others for themselves. "Do you still have the manufacturing plans? We may very well need more."

"Um, well, yes," Iza said. "They never asked for them back. Besides, we could just design more based on the ones we got. We may need to replace them if they get damaged or just wear out. You say we might require them for quite a while. I have some ideas for improvements if we need to make more."

"I think we should name them!" Xiang Xao Jia piped in. This was the OBGYN physician who was now engaged to Dr. Hosokawa. She was now generally considered to be part of the team and was therefore welcome to attend these monthly meetings. "Let's have a contest! Bring your suggestions next month!" This sort of random outburst was rather unusual in terms of the usual protocols of Adam moderating the input from those in attendance. But no one seemed to mind. It actually began a trend. Nearly everyone had realized by then that less rigidity and more camaraderie were going to be vital in the times that were coming. They would be confined in close quarters together for quite some time, after all. Things were getting real now, and they needed more than ever to decide how to interact together. They might have chosen to behave like a sort of military outfit. But instead, they took this chance to make themselves into more of a family-ish group of comrades. Naming robots was something a family might do.

"I like the sound of that!" Adam agreed. "Could be fun. Now, I was thinking I'd let you all set the agenda today with the time we have left here. Anybody with urgent matters they need to bring up, just raise your hands." Nearly everyone did. But Azaan Nazari got his hand up first. His facial expression also seemed the most distressed. "Go ahead, Azaan."

"I won't go into anything we discussed the last time we spoke," he said. "But I wanted to bring up something I ran across in my research. I hadn't given this any thought before, but I've learned to be a bit more audacious in my thinking recently."

"Oh dear!" Adam said. "Go on."

"Now, I know we are all Star Trek fans," he began, leading to a few low-level groans in the room. "Now listen," he continued. "I think all of you have taken enough physics to know that there truly is such a thing as antimatter and that it can be and has been produced in various accelerators for quite a few years now. Of course, Star Trek vessels have purported to use antimatter drives since the late 1960s. But theorists all agree that mixing matter and antimatter actually would produce all manner of energy and require very little weight for fuel. Various proposals for engine designs have popped up for years, but no one has had the money or resources to even consider building one."

"Including us!" Adam barked softly.

"Well, that's quite literally true," Azaan continued. "But thinking objectively, the main problem is collecting the antimatter itself. We couldn't build a supercollider that could make it for us with all the money and resources we have or ever will have. Even if some country's government would let us use theirs, it would take decades to make the five grams of antiprotons we'd need to get to the next star system over. And it would take another five if we wanted enough to get back. Now, actually, the design for such an engine wouldn't be terribly different from other nuclear engines. It's just the fuel that's the problem. But if we had it, combining antiprotons with hydrogen nuclei in an enclosed magnetic field, or rather a long shaft and nozzle with a magnetic field for shielding, could get us to a nearby star within as little as twenty years or even less." The room filled with what basically sounded like a chorus of raspberries on the audio recording. "Well, check the physics yourselves if you want to. Not only that, but if we control the reactions properly, we can ensure ourselves an equivalent of one G of acceleration force for, like, 75% of the trip, and the deceleration as well."

"I'm not sure where you're going with this," Adam said. "But I was really hoping that actual feasibility will be the governing principle of this task force."

"No wait, just suspend your disbelief for a little bit longer!" Azaan insisted. "So, really, the only problem is the antimatter, right? Right!"

Awamila Hayat, the team astrophysicist, now joined the discussion.

"I know what you're thinking!" she said. "You're going to point out that antiprotons are widely suspected to be present in the Van Allen belts around the Earth. Just remember that this hasn't been definitively confirmed to everyone's satisfaction. There are theories about how to collect it, and it would undoubtably be a massive undertaking to do so."

"Yes, I know," Azaan admitted. "And I also know that it's quite dangerous for living organisms to linger in the belt zones, with all the radiation and so forth. But I've actually taken the step of designing some magnetic containment vessels that could be deployed to try it, just to see if it could be done. I think they could actually work! We'd need to deploy a few hundred of these, I think. We'd give them their own little maneuvering rocket jets to help them putter around looking for anti-protons and then come back to a Shrike when they're full. I also think we could make them undetectable so no one on the ground would see what they were doing and get nervous about it. Besides, they'd be in the Van Allen belts where nothing else wants to be, who would even be looking for them? One final bonus is that we have a near monopoly in orbital shipping. We could sprinkle them around whenever a Shrike is deployed on other business. I can have a containment tank installed at my new lab to receive them when they come back to base. We can see how much we can sequester by the time the ship is ready. If it all works out, it should only take a few years."

"And if we get it wrong, we'll blow up most of the United States, won't we?" Adam said rather somberly.

"I suppose so," Azaan agreed meekly. "But antimatter has been isolated, and then destroyed on numerous occasions in accelerators in the past. It's just not publicized. The Earth is still here if I'm not mistaken. Besides, we do have some very powerful supercomputers around here to help us out. I really think this could be feasible. Plus, we wouldn't have to do some insane superspy coup to steal huge amounts of nuclear material for an atomic engine. We should probably put some distance between ourselves and this planet before we light this thing up. But like Adam says, I think if we don't do it, someone else will sometime in the future, and we'll be sitting here wishing it had been us. What's the harm in just giving it a try?"

"All kinds of harm, actually," Awamila answered. "But I suppose all forward leaps entail risk. I'm reminded of when petroleum was first developed and then atomic weapons. Everyone thought these things would cause the end of the world."

"They still might be," Adam said curtly. "But this time, it's an even stronger possibility than anything done in the past. But I guess I'm OK with exploring it as long as we watch our steps very, very carefully." No one else in the room could find the courage to disagree, at least on the recorded minutes.

Chapter 11

The Uplift Orbital Research Laboratory

From compiled meeting notes and personal notes from Nathaniel Floatingfeather…

To his own surprise, by mid-November of 2085, Adam felt his plans were sufficiently advanced to consider beginning construction of what he told the FAA would be called the "Uplift Orbital Research Facility." He wanted this unprecedented project to be completed in as short a time as possible, mainly so he would have more time to utilize it. The rest of the team indicated to him that they all felt the same way. All of the team members were still quite young, which was why they had been recruited in the first place. This was entirely on purpose. Only Dr. Hosokawa was over the age of thirty, and he was only 32. On the first Monday of October, Adam made the announcement to the entire task force that they were all expected to schedule themselves for flights aboard Uplift Shrikes as often as they were able. The recently purchased hotel disc would be fitted out to act as dormitories for those who needed to stay in orbit to better accomplish their assigned task force goals. Since the first assigned objective was to assemble the huge backbone of the new station, Manjeet and Gregor would jointly supervise this assignment from orbit. They would alternately stay aboard the hotel portion, or cycle back to Earth on Shrikes as their duties would dictate. But at least one of them would always be at the hotel or "small ring" site. Work would be completed using robot arms on the Shrikes or by the four construction robots already aboard the station. Since there was already an airlock in the docking bay in the center of the small ring, four EVA suits and tools would be brought up in case there was any need for more direct human participation.

Not missing out on her chance to follow up with her suggestion of naming the robots, Jia, as she had asked to be addressed, was quick to bring this up as soon as the robots were mentioned at the next task force meeting. When the topic was addressed, everyone was asked to write four names on a piece of paper and pass them to her. Jia took it upon herself to read them aloud and called for a vote. The group deliberated in a surprisingly lively fashion before the final names were assigned. None of the names seemed particularly serious, but obviously silly ones, like "Twinkle Toes" and 'Powderpuff," were thankfully discarded. The chosen names were slightly less silly and seemed rather maritime in nature. This seemed fitting for the construction phase of what they assumed would eventually be called a "ship." With some small degree of fanfare, the consensus was for the first bot to be named "Blackbeard," the second "Sinbad," the third "Popeye," and the fourth "Sparky." Sparky was somewhat of an outlier as far as names went, but it did seem appropriate. This poor fellow was to be designated as the robot assigned to the rear of the ship, near the engines. He would be the one assigned to maintain and repair the ship's engines, where preposterously high electrical voltages and even higher amounts of radiation would likely be present. Once posted in this hellish environment, he could never again return to the habitable areas. It seemed probable that he himself would inevitably become highly radioactive himself and, therefore, be unfitted to interact with his living crewmates ever again. If he ever broke down, he would simply be cast overboard, and a new Sparky would be built. Xao Jia volunteered to label the bots sometime in the future so she could tell who was whom when the crew started coming on board.

Recapping the work that lay immediately ahead, remember that this original "small ring" of pods that were part of the original hotel complex had a central solid disc-like area that contained a docking complex in the center. When one approached the hotel, this central portion appeared as though it might be entirely solid from the outside. But inside it was actually filled with a multitude of compartments containing a control room area as well as storage areas, tanks of fuel or water, heating and cooling equipment, and compressed breathing gases in even more pressurized tanks. In this center portion of the wheel, passengers and crew were subject to the worst spinning sensation to be experienced anywhere on the station. This lessened the further you got away from the center. Once you made it to the outer ring of pods, you could walk on flooring placed at the very outside of the ring. That's when you felt as though you might be bouncing along on the surface of the moon. Those unfortunate crew members who happened to be posted in the control center, or "bridge" if you prefer, tended to have the most trouble accommodating to their work environment. At least there was a large window looking towards the front of the station on the bridge. This could be darkened if desired if too much light was shining into the compartment. If you didn't want to see the Earth spinning around out the window for some reason, you could blot it out with the flick of a switch. Otherwise, you might need to find a motion sickness bag almost immediately. Fortunately, once you left the internal disc, the outer ring of habitable pods was quickly accessible via four spoke-like tubes radiating from the center disc. The spokes had elevators within them to allow guests to "ascend" to the outer ring. Paradoxically, the spin of the wheel made it feel like you were being pulled downward, so signs on the elevator doors told you to flip over so your feet pointed upward when you floated into the compartment. The whole assembly, turntable, spokes, and ring were kept in constant rotation to keep that low gravity field as constant as possible. That way at least passengers could eventually get used to it. In practice though, few guests really ever did.

In daily operation the docking complex in the center of the entire structure rotated right along with the rest of the station. Generally, when a Shrike approached with its cargo of supplies and hotel guests and wanted to dock, it would need to create a longitudinal spin to match the station's as it approached. This was done using the Shrikes maneuvering thrusters, those little RCS motors. The belly would be positioned "down" so that the bottom of the Shrike would face the "floor" of the docking bay. A combination of magnets and latches would secure the Shrike when it was in position with its landing gear retracted. It was then pulled forward into the bay, and an extendable airlock would be maneuvered out to seal up with the hatch in the bay of the Shrike, which would be opened on touchdown. The landing bay could actually accommodate two Shrikes at one time, but that was a tight fit. The outer habitation ring had four docking ports of its own as well if one didn't need to go through registration in the center reception area. These outer access airlocks were used mainly for transferring personnel or for bringing small packets of cargo inside the station. It was mandatory that everything that was brought inside was packaged in small enough bundles to pass through the four outer air locks as well as all internal passageways.

Now, for the Orbital Research Facility build, the task at hand was to attach the existing small wheel to a much larger frame structure aft of it that would ultimately become the station's literal spine. It was decided that the small ring's rotation would need to be arrested for at least a week while construction was in progress. In practical terms, this was a bit of a relief to the work crews since they didn't have to accustom themselves to the infernal rotation every time they were rotated from the ground to the station for work shifts. Shrikes could pull in nose first after opening their cargo bay doors but wouldn't need to match the station's rotation when they docked. The Shrikes would have a cargo container fitted tightly inside their cargo bays. Docking adapters were fitted in the cargo bays to allow the crews to move from the crew living spaces on their lower decks to either the cargo bay or into the station as needed. This would also work if the Shrike was docked to one of the air locks on the outer ring. In the main center docking bay back when the hotel was still in use, the Shrike would simply thrust forward to exit out the back side after everything was unloaded or reloaded with departing passengers or equipment.

With this new construction, a large hub would be assembled from pieces that would fit in Shrike cargo bays and bolted to the back of the disc area. This would cover up the rear docking port exit area completely. This assembly would introduce a rotating turntable-type plate that would serve as the mounting surface to which a frame for everything behind it would be anchored. The crew corridor connecting the forward wheel to everything aft would be built that would pass through the center of the turntable hub, then continue back towards the rear of the station. Three smaller access shafts containing ladders radiated from the center forward disc area in front of the turntable itself. These would themselves rotate to match the rotation of the small wheel. When reaching the front wheel from the rear of the station, one would have to grab a set of handholds for the corridor you wished to ascend which would be rotating when you arrived. Once your floating body adjusted to the spinning, you could then go down the ladder feet first towards the centrifugal forces until you reached the interior of the central disc. Once you recovered, you could then find your way to the actual elevators to access the outer ring of pods. If you wanted to go to the bridge or other inner disc areas, you would most likely want to take a meclizine tablet. One impactful consequence the addition of the central spine structure was that by attaching the new rotational turntable system, was that once you pulled a Shrike in through the front entrance, you would have to back it out again when you wanted to leave. This made things a bit trickier for the Shrike pilots. Fortunately, automatic piloting systems aboard the Shrikes would simplify things immensely. You just had to be really sure that the next Shrike wasn't coming up behind you. If you think all of this sounds bewilderingly complex, you aren't alone.

After the arrangements for the rotation of the small wheel were implemented, which was challenging enough by itself, it seemed fairly straightforward to begin extending the main structural truss system backward toward the rear of the station. This did, in fact, wind up being a much simpler process. Shrikes needed only to maneuver into position, open their doors, and use their robot arms to lift out the trusses and allow them to extend to their proper lengths. Sinbad, Popeye, and Blackbeard would take over from there, ensuring proper alignment of each new part and then welding all the pieces together. Power and communication lines, as well as conduits for heat, water, and atmospheric gases, were added next. This was also found to be a sensible place for storage tanks, antennas, and other equipment that didn't particularly need to take up interior space, which would unnecessarily crowd the crew. This process took only about two weeks to complete and was taken as a good omen by the design czars Manjeet and Gregor.

The next phase of the build was much more daunting though, and much, much more nerve-wracking for everyone involved. This was, of course, the construction of the so-called "Big Wheel" that was to be the main crew habitation area for many months or even years to come. All crew habitation pods, food production areas, recreation, medical spaces, and all other necessary activities were to occur in this wheel. According to the plan, the first step was the installation of yet another, and still larger type of turntable to allow rotation of this much larger rear wheel. This wheel at least had the advantage of not having a large docking bay running through its center to work around. There would be elevators, not ladders, in these four radiating spokes. When you finished floating down the central corridor from the forward wheel, you would arrive in a central vestibule to find four elevator doors rotating around you. Once you chose the elevator you preferred, you would grab onto one of the handrails by that door and synchronize your body to the rotation of the wall, as with the smaller wheel, then push the up button. Once aboard the elevator, your motion would stabilize, and you would rotate yourself to get your feet in the direction of what would become the floor as you were raised to the exterior ring of habitation pods. After some experience, you just got used to getting on the elevator upside down. All of this shifting of orientation would undoubtedly take a lot of getting used to, but there would be plenty of time allotted to learn one's way around. Adam once said that learning to live in these conditions was rather like finding oneself in the middle of an Escher painting. He came to understand that in zero gravity conditions, an Escher-designed room actually would make perfect sense. Even so, crew members would go to great lengths to arrange their schedules to avoid needing to travel the maze to the forward wheel as much as possible. Fortunately, in the long run, most crew time would be spent in the big wheel. That was where you could at least tell the difference between up and down. Your body and mind would stay healthier if you spent most of your time in this main wheel.

The first step in the construction of the big wheel was the assembly of a somewhat more elaborate spinning mount. The impressive structure required an inordinate number of Shrike missions to transport all the cargo bay sized pieces that the bots had to put together. When this process was finally completed, an even more elaborate truss system than what had been built before was extended to allow for future construction of the rearward engineering structures. Automated Shrike piloting made the required maneuvers doable, but a good deal of patience remained necessary for the construction crews. This new portion of the frame went backward for about 150 further meters. Once this framing was completed, assembly of the four hollow spokes commenced. The elevators, as well as a small bevy of ducts, pipes, and wiring, took another few weeks to complete. Three huge circles of connecting pods, the true heart of the ship, could then be assembled to make the main habitation zone. The key to the construction strategy was to make sure everything brought on board for the build was packaged in such a way that everything needed for the interior would fit through every door and passageway on the station.

In the rear areas behind the big wheel, two other habitable structures would be constructed. Sharon had suggested that a large Zero-G space be built that was intended to be used as a huge arboretum, observation platform, and free-floating recreation area. She observed that trees probably would not be particularly bothered by Zero-G; at least, this had been true for the saplings that had been grown in orbit so far. They would also be small when brought aboard, and it would be interesting to see how they fared over time. At the very least, they were certain to convert a good deal of accumulated CO2 back into oxygen. Assembling this large open area required solid frame rods to be assembled, resulting in open spaces to be filled with thick plates of plexiglass. It would ultimately take more than a few Shrike Missions to lift this all into place. But the task force was quite adamant that this feature be included, if for no other reason than to aide in the well-being of the crew. On the opposite side of the aft frame structure, and also connected with corridor tubes, in the area below the arboretum a Physics and Astronomy lab was to be built. This was Awamila's bailiwick and would obviously house the telescope.

Additionally, various sensors and equipment to measure the particle densities, gravity fields, and general environment outside the station would be installed. Beyond the arboretum and physics lab, a huge radiation and heat shield will be constructed to keep the habitable portions of the station fit for human occupation. All the other portions of the frame would soon be crammed with various pumps, heaters, wiring and other equipment. Finally, the theorized propulsion systems would eventually be constructed back beyond the huge shield. More conventionally, the huge but ultralight solar sails were to be mounted to the main frame just behind the front habitation wheel, as were the solar power panels.

By the last week of November 2085, all attention was on the installation of that first bit of truss onto the back of an unused orbital hotel. Objectively and in retrospect, this feat was really a marvel of rapid design, engineering, and manufacture. Again, the advent of supercomputing as a facilitator of design was a monumental advance. True, this was not a complete novelty by the 2080s, but by the time Junaki Sato was utilizing these same processes to assist her husband and Uplift Aeronautics in general. Designers could, therefore, envision, design, and implement nearly any sort of needed structure within a matter of weeks. Last-minute ideas and changes could be added at will, at least unless a complete redesign was found to be needed. In the entire orbital facility build, full redesigns actually only occurred a handful of times. The supercomputers generally could inform the builders of needs they never even knew they had. The structural trusses were designed from the outset to be broken down into collapsible pieces that could be loaded into Shrike cargo holds and assembled in zero-G and at very low temperatures. These types of general principles were longstanding, having been done since the 1980's and 90's by NASA. These prior experiences were easily transmittable into computer simulations. Some cargos were just single thick linear bars, others connecting joints, others cross pieces and corners. The large turntable structures were made as pieces of a circle that the robots could bolt or weld together. One of the hardest steps involved attaching connections to the preexisting wheel already in orbit. The rotation of the wheel had to be completely nullified in order to allow the new pieces to be bolted and welded to the circular hub. The wheel also needed to be depressurized and was therefore uninhabitable for nearly a month. All construction personnel up to that time were therefore confined to the Shrikes. Eventually though, crews and crew chiefs were able to inhabit the comparatively more comfortable hotel suites for the remainder of the build.

One thing that Manjeet had realized fairly early on would be addressed with the assembly of the big habitation wheel. This concerned what would happen at times when the ship was accelerating forward at significant speed. At such times, the force of acceleration would propel the occupants to the back of the ring, not outwardly to the sides. The center hub of the small wheel, which also had a fair number of storage areas and equipment, would have the same problem. The small ring of course contained the main bridge area, and it certainly seemed necessary to make that as habitable as possible regardless of the G-forces at play at any given moment. Plans were instigated to convert these spaces to accommodate shifts in inertial forces now during general construction since there were crews available to do it. It was already known that zero-G conditions would generally rule in the hub areas for much of the time when the wheel was spinning. With regard to the trusses, it was deduced early on thanks to the computer simulations that the structures all over the station would need to tolerate truly massive amounts of stress. Although changes in velocity would be as gradual as possible, one could not get away from the fact that when engines placed at the rear of the vehicle were engaged, there would be a possibility that the engine modules would simply overrun whatever was ahead of them. It would be like having a super powerful engine at the back of the train, flattening all the cars ahead of it when brought to full steam. This is not something one generally thinks about when watching science fiction spacecraft in movies. But for designers, it matters a great deal. If such a thing as an antimatter drive could be designed and built, it would be somewhat challenging to build up speed gradually. Acceleration might be sudden and extremely forceful. Therefore, the trusses were being built with carbon fiber technology and infused with as much Titanium as Uplift could get their hands on. According to the simulations, all of this should work, but the number of cross-connections and supports made the trusses look like pieces of modern art. It all seemed to work in simulations. Thankfully, they proved relatively easy to assemble as well, at least from the robot's perspective.

By late January 2086, and the use of some 200 Shrike missions, the major truss structures back to the connecting turntable for the second wheel was complete. Next came the four spokes and elevators for the second, major wheel. As a reminder, there were to be a total of 300 pods. In addition to the actual 100 pod habitation ring, two 100 pod corridors would be placed to allow crew access in two different inertial modes, one for circular rotation, and one for forward acceleration. These would run 90 degrees in relation to each other to make an L-shaped trio of tubular corridors. The habitation pods would be the angle of the L. These 100 pods, you may recall, would be used for crew quarters, water storage complete with aquatic habitats, office space, medical facilities, and recreation spaces. The habitation modules would have doors (hatches really) leading to either corridor which would be used depending on where the acceleration was applied. This circular trio of tubes would then be enclosed with a solid outer shell covering to assist with thermal conservation and protection from micrometeorites and radiation. The spokes would connect to the service corridor rings with their internal elevators. The G-force environment within the elevators would make each trip in them a constant challenge, but they would get you where you needed to go. Once they were installed by the end of February 2086, a total of two months and over one hundred Shrikes flying at least three missions apiece had been needed to connect all of the pods in their assigned locations.

These pods were significantly larger than the ones used for the hotel build and were generally thicker in girth. Their inflatable skins were also thicker, and they had more attachment points for gear and ducting, both inside and outside. Once attached to a robot arm, they were lifted out of the cargo bay, inflated at low pressure with Oxygen, then connected to the last pod to be installed. At the end of each spoke, a sort of six-way cross connector had been installed. This allowed for an elevator terminal connection, two habitation ring connections, and a connection to an outer docking bay for any visiting Shrike. The other corridor- forming rings would connect to the habitation ring at intervals, but also to this hub. The connectors were installed at the ends of the spokes first. Once the pod was installed at low pressure, it was then fully pressurized with an Oxygen-Nitrogen mix at atmospheric pressure. Each pod had a connector ring on either end with a thin but rigid hatch that would be removed after the seal was confirmed. The hatch would be brought inside the new pod and be moved forward to be used again when the next one was added. This eliminated the need for three hundred-plus hatches in the three rings. Each habitation pod was left with hatches that separated them from each of the two service corridors, though only one corridor was actually accessible at any one time. There were also hatches between the living area and the bedroom area. The bathrooms were equipped with a folding privacy screen, as an entire hatch assembly was deemed to be a bit excessive. All of these mechanisms had been thoroughly tested and found to be very soundly designed. No atmospheric leaks at the joints or elsewhere were ever experienced during or after they were assembled.

This is not to say that this novel building process was anything but extremely labor-intensive. Usually, one Shrike was assigned to an airlock near one of the four spokes and was timed to ensure a safe distance from the other three. On arrival, each would inflate and attach one pod in the chain, then return to Uplift in sequence. Usually, these launches were done only during the daytime (in New Mexico) to avoid totally exhausting the crews. Three groups of four Shrikes would launch together. One group worked on the habitation pods, and the other two each dedicated themselves to each of the two service corridor rings. On many days, two sets of these four Shrike relays could be successfully repeated. This represented a very large commitment of fuel, crews, and support personnel. It would eventually be revealed that Adam Thorne was providing much of the funding for the build with his own money. As he had often demonstrated, it was as if money wasn't particularly important to him.

Besides utilizing Adam's personal wealth, Uplift was accepting a large amount of research money from a variety of firms. The largest sums were coming from medical research foundations, who were anxious to see if manufacturing in zero-G would optimize the production of new medications and technologies. NASA itself provided many millions of dollars as well, and the results of their own research contributed to some of the new techniques of construction that were being pioneered. Scientists there were envisioning future projects that would be based on this station, although they certainly never suspected that Adam had any unannounced intentions about plans to possibly make his new structure anything but Earth-based.

The big enabling factor to proceeding with the project was the greatly reduced cost of both manufacturing the materials in general, and also the reduced cost of shooting them into orbit. Adam always compared this period of aerospace expansion to the explosion of shipbuilding with the age of exploration in the early 16th century. Once it was found to be feasible to travel long distances over water, wooden shipbuilding and design expanded exponentially. Adam couldn't wait to find out how humanity might be enriched if this project was even marginally successful. Then again, if he was successful, he literally wouldn't be able to participate in the success. The point was that Uplift was by no means bankrupted by building this "station." He was confident that if he did leave, Uplift would go on profiting without him. That must have been reassuring.

In summary, by the beginning of April 2086, the original hotel ring had been attached to an extended framework upon which the other habitable areas were to be built. Connecting passageways were placed, as well as an array of storage tanks for liquid breathing oxygen and nitrogen, wiring for electrical systems, heating and circulation pumps, and communications equipment installed, or existing equipment improved. Also, the large infrastructure for the main habitation ring was now in place. The next major step was the installation of the large solar arrays. These were mounted to the central truss frame about halfway between the first and second wheels. They could be rotated along the vertical axis of their mounting poles to help assure maximum access to sunlight. There had been solar panels on the original ring, but these were small ones that were attached to each individual pod. The batteries they charged were in the central hub structure. The systems were expanded greatly, and soon the entire station had access to electricity. Tanks were also mounted that would serve the various reaction control jets that were placed strategically on both the frame system and on the exterior of about two dozen of the pods. Since there was a huge amount of exterior work to be done on the exterior of the pods, these were not yet operating. This work was up to the four constantly active bots, which greatly reduced the manpower that would have otherwise been required. Exterior heating and cooling lines would be placed to supplement the much more numerous interior ones. Sensor arrays were also to be used to monitor the skin condition of the pods. Each of the habitable pods has a total of six circular windows. There were also mounting brackets to accept the overlying external skin plates that would eventually cover the entire three tube system to make one solid disc. A layer of expandable insulating foam would be injected between the outer skin layer and the exterior of the pods for increased protection from the space environment as the work progressed. Openings would remain for the observation windows. Those much-abused bots would have their work cut out for them.

After the pods themselves were all in place, the next phase of construction began. This would perhaps be the greatest phase of the build in terms of concerted effort. Adam, Mattie, and Mangeet had realized from their earliest commitment to the project that the most vital commodity of this project would be water. Water was vital to every major function of the station, be it for human consumption, fuel production, food production, or to cool the inconceivable heat generated by the propulsion systems. Adam had already proved a Shrike could lift the stuff up in amounts of about 20 metric tons a mission. Conveniently, one metric ton was equal to 1 cubic meter of water. Since the habitation ring would have 100 individual pods, some of which were linked with open hatches between them, Mangeet and the other designers were tasked with allocating space to the various requested needs of all the various departments. It was decided that six out of every ten habitation pods would hold water, with the other four going to living and working space. This would give forty pods to individual and family living space, medical facilities, and common areas. The remaining sixty had to be filled with H20, and preferably loaded from the outside before the covering external layer was installed. These designated water pods had ports exposed to open space to allow them to be filled directly from the Shrike's tanks. First, the pods would need to be connected to a heat source, in this case electric internal heating wires, so that the water would not freeze when pumped into the pods. Also, a small antechamber was installed from the inside of each pod. They were accessible from the service corridors. They would house environmental controls, a circulating and aerating pump, and an access hatch into the water tank that would allow for a diver to enter it to perform maintenance and to feed and/or catch whatever fish might be in the tank. Other features, such as water filters and habitat augmentation for the fish would be added later. Also, there would be about equal amounts of fresh and salt water.

Bringing all of this water aboard was one of the only tasks where a full two thirds of Uplift's current fleet of over 140 Shrikes were employed full time for over a month. They flew around the clock. Bringing all of this water to the Uplift facility in the first place was a staggering endeavor in its own right. Fleets of tanker trucks were hired to bring in water from the Caribbean, Baja California and the West Coast. Fresh water came mostly from the aquifer below Truth or Consequences. All of it was launched from the main Uplift facility. Manjeet and Mattie had worked hard on deciding how to distribute all this weight around the ring, as well as in several supplemental tanks mounted elsewhere on the frame. In all, about 800 metric tons of liquid water were to be moved to Earth orbit. And this was just a preliminary figure. More would undoubtedly be needed. Later phases of construction would see delivery of equipment designed for hydrolysis of water into Hydrogen and Oxygen, as well as other units for combining them back into the water. Other tanks external to the ring would store Nitrogen and monitor and adjust the content of the breathable atmosphere in the habitable areas. The two service corridors of the big rings would house much of this equipment. This also applied to equipment for removing CO_2 from the air and directing it to areas populated with vegetation or simply dumping it overboard. Atmospheric moisture would also be condensed and sent back to one of the water reservoirs. Every computation made by the engineers showed that barring some major leak or other catastrophe, there should be ample water for a crew of 20 to last for at least one hundred years. This didn't even account for the reclaiming of water from bodily waste and other uses, which would also be done. Solid waste would go to fertilize the plant life.

In addition to the water itself, most Shrike cargos included some of the equipment for the meticulous management of $H2O$. The station truly became a giant hive of activity at this stage, and this frenzy never really alleviated until the entire ship was nearly completed. Shrike crews always consisted of at least four people, and for the most part, when they were up, they all stayed on board their Shrikes. Obviously, there were never more than maybe ten Shrikes around the station at one time, and they generally stayed aloft only to discharge their cargo before returning straight back to Uplift. Back at the base, management did need to deal with some complaints from the local population, who suddenly found themselves living next to one of the busiest airports in the world. Uplift announced that this extreme traffic would only last for a few weeks before returning to the usual low roar they had been used to all these years. They were lucky that Shrike flights relied on jet power until they reached the Caribbean Sea. Only then did they turn toward space and go supersonic with their rocket engines. A portion of the flights, however, and therefore the crews, served for short periods aboard the station to help install all of the pumps, plumbing, wiring, and innumerable other things required to make the major systems operable. The docking ports were quickly made functional, and these orbital construction specialists generally would go back to their ships to eat and sleep until all of the hardware they had brought up had been installed. A few of the more fortunate key personnel were boarded in the more commodious old hotel rooms in the small ring. Once all the water had been uploaded, Blackbeard, Sinbad, Popeye, and Sparky started working on the exterior shell components using materials handed to them from innumerable robotic arms. Shrikes were always lined up to hand them out. The bots worked around the clock, except for the short periods when they needed to return to their charging stations once or twice a day.

During these enormously hectic days, those who had time to notice did notice that Adam and Mattie seemed to have chosen to work together constantly. If they were crewing a Shrike, they took turns at the controls. If they stayed aboard the station, they stayed together. They assigned themselves a suite pod on the small ring once it was again able to be rotated. Once communication with the ground was dependable from the station, they were aloft far more than on the ground. Manjeet and Juni also seemed to have decided to emulate this behavior, although they did insist that they have weekends off back on the Earth's surface. The other task force personnel only came up if they were specifically needed. Generally speaking, Gregor Levinski and Iza Mazurkiewicz were really the only other key people who needed to be constantly aboard at this stage of the project. Iza had never been in space before but was having the time of her life working with "her" robots. She arranged early on to have them brought back aboard through the air locks in order to apply some decals she had had printed up. These had their names on sticky vinyl labels which she put on their right upper chest areas. Popeye had an anchor on his forearm, Blackbeard got an analog bit of facial hair, and Sinbad got an eye patch. Sparky was soon adorned with a variety of cartoon lightning bolts in various locations. She also fitted them with small speakers and a transmitter on the front of their "head" areas. They each also had two cameras that resembled eyes so operators on board the station could see what the bots were working on. All of this was connected to Juni's computer complex on board. The bots could now communicate with their operators, via Wi-Fi or in person, using the voices of their assumed personas. These emulated whatever the computer thought those voices should sound like. Popeye was the only one with a well-documented voice, and he was quite hilarious. People soon tried not to talk to him if they needed to really concentrate on what they were doing. It was too distracting. It seemed a pity that these four metallic crewmates were hardly ever inside.

And on the inside, enormous amounts of nonmetallic human work would be ongoing and difficult for months to come. Although both wheels could now be made to spin, it became apparent that some of the work would often be easier in zero-G. For example, when Sharon Layton made her first flight, she brought up a crate of artificial rocks to place in the first pod assigned to serve as a habitat for saltwater fish. She soon realized it was far easier to maneuver these bundles around when everything was floating weightless. Trying to coax bulky rocks into a tank of weightless water was another proposition, though. She wound up leaving them in the vestibule in front of the tank's access hatch, then dropping them in one at a time whenever the ring was spinning. Most everyone made similar decisions when bringing heavier equipment aboard. In point of fact, nearly everything designed for use on the station was produced with the minimum necessary mass. Mass would translate into weight when any sort of gravitational acceleration was applied later on. The less mass any sort of future engine had to push around, the better. It was always best to lighten the ship whenever that was possible. With experience, the builders would choose to move heavier items simply by using rather conventional dollies. For now, though, zero gravity was a clear and useful asset.

All of this intense effort had led to the point where, by the end of August 2086, the station had functional heat and lighting, full water stores, functional communications and attitude control. Adam then gave the order, and the two independent habitation rings began to rotate on a fairly permanent basis. The outer shell of the larger ring wasn't yet complete, but the bots never once verbalized any objection and kept right on going with the exterior work. Workers could now begin to fully fill the station with the endless myriad of components that would be required to make it truly habitable. Hopefully this habitation would be "in perpetuity" as the Uplift task force had prescribed. These components were broken down according to departments and were assigned to various members of the crew to develop. In many cases, the personnel themselves were in fact the components. It therefore seems more expeditious to chronicle the various department's activities individually. Since these people would wind up staying on board once the building was complete, they all needed to put their best feet forward to make sure the mission was as successful as possible and their lives would be tolerably comfortable. A list of devoted subsections for each component departments' contributions therefore seems to be most appropriate. These individual contributions will be combined into the next chapter in order to maintain a more conventional format. The information presented was compiled using records of work orders, recorded communications and recollections of the workers from the various departments. To be honest, some future biographer may find it most appropriate to write fuller accounts of each of the Adam's task force members individually. The ultimate achievements made by this group were only made possible because all of them were working together as a unit. Hopefully I will be able to convey the sense of urgency that was experienced by those participating in the station build. It was remarkable to observe as it all occurred.

Chapter 12
It Takes Villagers to Build a Village

Part 1: Supercomputing with Juni

It had by now been well established amongst the crew that utilizing state-of-the-art computer technology was an absolute necessity when attempting to fly about in space. Spacecraft designers from the earliest days have needed whatever assistance that was currently available to assist in the various aspects of spaceflight. One is immediately reminded of what was then the astonishing computer system placed aboard the Apollo command modules. Guidance and navigation were vital simply to achieve proper orbits, velocities and rendezvous approach strategies. Every protocol and vital task required had to be punched in by hand into a small keyboard, with a specified numeric code for each and every operation. Within ten years of the end of the moon program, even pocket calculators were far smaller and much more useful than any command or lunar module computer. During Apollo, NASA management used to brag that their computers could "fit in a single room!" and it was a very large room indeed. By the 2080s the computing equipment Juni was bringing aboard the orbital station also took up a relatively large overall area. The difference was that the newer equipment could contain pretty much all of the data and information available in the United States in the later part of the 21st century. This was very nearly equivalent to all of the information available on the Library of Congress servers, the Thorne Engineering Library, and even more. Navigation computers were placed in the storage areas near the bridge, near the center of the smaller ring. This complex was able to help compute ship velocities and plot a course to nearly any destination requested in the entire galaxy if requested to do so. In actual use, there was always just one destination in mind, and that never changed after it was decided upon. The ship was also equipped with radar and sensor arrays which could detect any object larger than a baseball at a distance of about 10,000 kilometers. If detected, the computer could automatically calculate maneuvers necessary to avoid that object and then restore the prior course by itself without ever consulting anyone. If Juni felt the need to consult, though, she certainly could do so. She decided to make it a point to personally ensure anyone on board the ship could use the equipment as well as she could. She was the rare sort of instructor who didn't simply assume that "Everybody knows that!" which is a common assumption among computer engineers. Once she had moved aboard permanently, she spent most of her days teaching how to use the systems to anyone who had time to listen. Since computers are really the heart, soul and brains of any spacecraft, knowing how to use them efficiently was absolutely essential. Surprisingly, Adam Thorne later admitted that of all the crew, he had had the least prior experience with computer system protocols. He had spent most of his youth and even his college days attending to aircraft and spacecraft dynamics and design. He was the sort of person who designed with intuition and a sort of inner vision. He had seldom fully utilized computers to complete his assignments. His professors tended to discourage computer use anyway.

Also unusual for engineering geeks, Adam also was never much of a computer gaming type. He dabbled a little on those rare occasions when he had social time with friends. Mainly, though, electronic games just kind of bored him. Juni and Manjeet were very much the opposite. Much of their time living together and then courting was devoted to computer gaming. Juni had promised early on that she would make computer-based entertainment a very accessible commodity on board the new orbital habitat. She tried very hard to make good on that promise. She had engineered and helped program a series of servers that held as much internet content as she could legally obtain, and this included video games. Educational tools that could be used to teach classes in everything from kindergarten to a plethora of PhDs were uploaded diligently into the memory. Earth geology and geography, star maps, art, music, literature, and every other topic the internet could suggest were included. Each habitation pod had a connection capability for each crew member's personal use. The hardware for these data banks was placed into one of the auxiliary access tunnels and connected to the general power supply. In this case, this power initially came from huge solar panels that were deployed fairly early in the project. Cooling lines were also employed to prevent the systems from overheating. Juni was pleased that these services were utilized pretty much universally as soon as crews began staying aboard the station. Another one of her promised services was the two interconnecting pods that she had fitted with surrounding flat video screens that would be used to create visual experiences for the crew. Usually each pod was used separately, although the two pods were connected with a hatch that could be left open if more space was needed for some reason, say, a social gathering. These could be reserved in advance and could provide the users with an almost unlimited variety of visual experiences, from a simulated bike ride in the Alps to a concert by your favorite band. Again, this wasn't a Star Trek-level holodeck, but Juni felt that making this technology available might be a good way to keep the crew connected to their own humanity and the planet they had left behind. It seemed to her that it might be a good teaching tool as well for any future users who might possibly be born off the planet. Also necessary for crew morale, Juni also made sure that cell phone communications would be available to everyone anywhere inside the station. This was something everyone had become used to back at home. Therefore, it was deemed absolutely essential for everyone to be able to communicate with each other and with the people on the ground whenever necessary.

Juni's next project had to do with the clinical assistance programming the medical people had requested. An entirely separate but still connected set of servers had been ordered and programed for the use of those crew members. A suite of five pods was currently envisioned exclusively for medical purposes, and the servers for their use were located in a service corridor adjacent to those pods. The clinical database included an encyclopedia of treatments, diagnosis protocols, pharmaceutical information, surgical standards and all medical texts that were endorsed by the American Medical Association. She had even taken the initiative to download protocols for use in robotic surgery. Iza had informed her that she intended to have two of these robo-surgeons fabricated for the Doctors to utilize in case of any need for assistants, or if the surgeons themselves were incapacitated. Who could predict whether or not the surgeons would eventually ever need surgery themselves? Any objective thinker soon realized that the gamut of possible medical requirements aboard a prolonged flight were very nearly infinite. Juni tried her hardest to try to prepare the computers for whatever scenarios might occur when no assistance was available from the ground. She and Dr. Hosokawa were in daily contact, although no medical personnel had yet come aboard.

It also defaulted to Juni to oversee the communications not only on board the station, but communication from space to ground as well. She had to consider that if any of these nebulous plans for leaving Earth orbit ever came to pass, the ground might be very far away indeed. As a starting point, a very large dish was ordered and installed at the front of the station and hung below the first wheel to enable radio communications as was usual in orbit. A communications station was wired into the bridge compartment. Additionally, and redundantly, an external array of smaller laser communications reception and transmission equipment was affixed to three places on the frame trellis. This would theoretically come into play exclusively if the station ever actually left Earth orbit. She had actually already commissioned a set of ten mini satellites that would be powered by small nuclear batteries that could receive, amplify, and relay laser encrypted data back to Earth at great distances, and hopefully receive anything the Earth attempted to send up to them. This would hopefully keep the ship in contact with the Earth at somewhat more timely speeds than plain old radio signals without relays. When she began to dabble in developing this capability, she immediately realized this was a more complicated proposition than she initially supposed. For example, if the ship merely released these satellites at regular intervals, the satellites themselves would also be traveling at extreme speeds and basically stay right there with the ship as it traveled along. Therefore, they would themselves need some sort of motive force to slow down, optimally to a stationary position in space for the rest of eternity. She decided to design a catapult apparatus to fling them backward towards the Earth, as well as a small rocket propulsion module to at least try to reverse the forward momentum. There were also small sail-like structures that she hoped might get caught up in the stream of energy emitted by the ship's engine. But the relay device would need to have small RCS jets of their own that could catch the contrail stream further away from the ship where it would be less intense. She soon realized that all of this might not work at all, but everyone would expect that it should be tried. All of these ideas were under development by a special team she had assembled and wouldn't even be installed unless someone decided the ship would actually leave Earth orbit. This team's work was not openly advertised much back at Uplift.

One final area where Juni had responsibility was to assist the astrophysics mission efforts to be performed on the station. Obviously, she worked closely with Awamila Hayat, who was herself working on a separate structure behind the big ring where she would place a laboratory compartment unlike any that had ever been conceived. A vast array of monitoring equipment, as well as the most sensitive telescope that could be fabricated, donated, or otherwise attained for the station. Apparently, word was going out among the scientific community about the need for such a scope. Awamila felt hopeful that someone would step up and donate one. Since no one had been authorized to leak any sort of information about plans to possibly leave Earth orbit, no one, not even NASA, would really see a need to donate a top-of-the-line telescope. The James Webb orbiting telescope had been deployed some sixty-five years before. It was no longer viable, obviously, but no one outside of government agencies had found a financially sound reason to follow up on its discoveries. If this Uplift station ever did leave Earth orbit however and found itself actually traveling outside Earth's very solar system, going without a top-notch telescope would be a virtual crime against humanity. Human presence in interstellar space made this orders of magnitude more imperative. The point was that Mila's dedicated lab space, located in the rear of the habitable frame structure, would be in obvious need of supercomputing hardware. Juni had already approved the design for this by the time the lab's construction began and was helping Mila try to obtain a telescope throughout the entire building period. The trellis frame for this compartment as well as the arboretum also being manufactured in Grants and would be lifted and deployed as soon as FAA approval was granted. The relatively large greenhouse structure would be affixed to the other side of the frame from the lab. The two compartments would be accessible from the central main corridor and would share the same power supply. Completing these two venues would take a few months at least. The main assumption was that this elusive telescope would need to be transported in a Shrike. If so, it would probably be in pieces and would need to be assembled inside the lab structure when delivered. Only then the surrounding windows and bulkheads would be completed around it. It seemed impossible for any descent telescope to be broken down and then carried in through hatches and reassembled. But Mila didn't know if such a telescope existed. She had to search around and find out. All these sorts of arrangements were Awamila's responsibility during the build. There were no easy jobs when serving on the Sustained Orbital Facility task force.

Part 2: Home Décor with Manjeet and Gregor

Mangeet and Juni had moved into one of the original hotel room pods as soon as the wheels began rotating. Both of them were by this time seasoned Shrike pilots and were used to weightlessness. But they both wanted to get used to the feel of this angular momentum-ized environment as soon as they possibly could, if only to see how they could function. To their disappointment, they both began to feel the effects of motion sickness within about twenty-four hours of moving in. They could therefore understand then why the enthusiasm for tourist visits to the hotel had dropped to near zero in such a short time. They did bring up enough personal effects to their room to enable them to stay aboard almost constantly, although their pod was used basically as a storage locker. They did manage to sleep in the sleeping bags slung there from time to time. Whenever they could help it though, they would spend time in the zero-G but spinning central hub of the small wheel. Since they were spinning right along with it, they didn't really notice as long as they blacked out any windows in the room. They were distracted with motion sickness by their work when in other areas though. This was generally the case for many others working on the build, who were even known to steal bunks in visiting Shrikes if these were available for sleeping periods. These unpleasant experiences motivated Manjeet to press on as quickly as he could with outfitting the habitation pods in the big wheel, as working there couldn't help but be more pleasant. Gregor was also assigned this duty, along with all the personnel who worked from the Shrikes, to assist them. Fortunately, there were quite a lot of them. Juni had a fair number of people under her as well, electricians mainly. But a good deal of her work was hers alone, with close communication contact with her people on the ground. She had about everything she was responsible for in place aboard the station in around four weeks. This was enormously useful, rather, very much indispensable to everyone else on the station.

When the big ring infrastructure was completed and sealed up, it was immediately seen to be vital that the living quarters areas should be first to get fitted out. Gregor Levinski, in addition to having designed air purification and heating systems for this part of the station, had taken a special interest in how these compact little apartment-like spaces should be built and installed. With the unspoken understanding that all members of the task force group intended to join the crew, at least initially, he had sent out e-mails with a catalog of the Ikea-like furniture pieces he was intending to have built. For example, dressers, which were really just wooden lockers with drawers and cabinets that latched tightly when closed, were at least available in a variety of styles. There were beds that could be folded up to make the compartments roomier when not needed. The common name for these was and remains, Murphy beds. These could also be ordered either in a bunk configuration or a single queen-sized option. These were also equipped with head and footboards that had hooks for slinging hammocks suspended over the mattresses in zero-G periods if needed. Each "family unit," whether there was just one person or several people per family, was to be allotted two pods, one as a combination living room and kitchen and one as a bedroom for sleeping as many individuals as the family included. The living pod kitchenettes would each be equipped with a refrigerator, stove, microwave oven, and clean-up area. The "clean-up" area included a sink, which was not quite a sink but had a water dispensing faucet and a deep tray with a vacuum feature that would suck up water or cleaning fluid and pump it down the drain. If G forces were sufficient, then it behaved in a more sink-like fashion. It was therefore necessary that the bottom of this cleaning tray point "downwards" or in the direction of the prevailing downward force. This problem applied to everything in the living pods. As Manjeet had envisioned, most of the furniture was to be mounted to what was basically a fiberglass inner container that fit inside the cylindrical pods. The floors were flat to erase the cylindrical curve that would have been the floor if the inserted "box" weren't there. There was no ceiling, however, so the upper curve allowed for increased headroom. These boxes were connected with bearings and wheels that fit into aluminum tracks encircling the inside of the pods. This would allow the entire box structure to be rotated to accommodate whatever forces were being exerted at any given time. Manjeet had actually had motors manufactured that would allow the occupants to rotate their compartments themselves as desired using a wall mounted switch. But generally, all habitable areas could be controlled in unison from the bridge and could all be adjusted in unison. There was an exit from each of the pods that would lead to one of the two service corridors. The corridor chosen would depend on whether the wheel was spinning or not. If the ship were accelerating forward, then no spinning was necessary. The force would push you toward the stern of the ship, and the corridor to the side was used. You would then step sideways into your pod. If spinning, the corridor to the inside was used. To change from one mode of force to the other, the room would rotate to place it in the proper orientation. The design team was really clever to have deduced this Escher-like solution, but it did make assembling this puzzle a lot more challenging.

Gregor had assisted in designing interior materials so that they could be packaged in pallets that would fit aboard the Shrikes in easily moved bundles. After docking, the pallets would be passed through airlocks and could fit in the elevators as well. It was generally preferred to use the air locks at the periphery of the big wheel. The Shrikes were capable of easily docking to these even when the wheel was spinning. This was found to be a bit difficult for even the most experienced pilot to do manually though. On the whole, the construction people found it easiest to assemble the furniture kits with a bit of artificial gravity, but it could be done in weightlessness if that were necessary. Even the kitchenette storage cabinets were part of the kits, but appliances were separate. Full-sized refrigerators/freezers, stove tops, and microwaves were all shipped up independently. As described, there was the sink-like set up to install as well. Gregor had eventually realized that a pressurized water-using dishwasher was also something he could design and have built. These were also included in every kitchenette, although this necessitated that a larger water filtering and waste collecting method needed to be designed. He had already planned for an on-board processing plant to be placed in one of the service corridors, and he made the necessary modifications for these changes. He used the types of these used on large cruise ships as his models but found that smaller, lighter pumps and machinery would suffice for the much smaller crew. The stove was electric, but it was nice to remember that the station would have an atmosphere of Oxygen and Nitrogen, so the fire danger was somewhat lessened. This was already in place and in use, and it seemed to be working without problems.

At the far end of each bedroom pod, there would also be bathing and toileting equipment. The toilets would be of the same design already used in zero-G situations on space stations or small transport spacecraft. This included Shrikes, although Shrike "facilities" were quite a bit smaller. Both functions depended on suction pumps to either suck up the human waste and send it to the processing tanks or to suck up the shower water and do the same thing. There were electrical outlets to allow for things like blow dryers and electric shavers. There was also a vacuum with a hose to suck up any floating material that might be generated in zero-G. Towels and toiletries would be stored in a moderate-sized cabinet in the bathroom. The privy area was separated from the sleeping area and had a sliding screen to separate the two zones. This would provide a small bit of privacy if the occupants wanted it. One other obvious question was how the crew members would do their laundry. In previous prolonged space stays in places like the ISS and even on Shrikes, the crew wore disposable garments which were tossed as soon as they became unpleasant in one way or another. The crew member would then go fetch another garment from storage and put the old one in a plastic bag to either eject out into space or bring home for the NASA dumpster. In this case, that would involve a lot of trash to dump and a lot of jumpsuits to 3D print over the potentially long period of the flight. Jumpsuits were again planned to be the only garment one could wear, but a couple of pods were designated as laundromats. Washers and driers were designed and also built which would function in zero-G quite effectively. Crew family units would be assigned certain laundry days once weekly, although ongoing work scheduling was not something that had yet been considered. During construction, before these facilities were installed, everyone just continued to toss out their dirty clothes as they always had. One thing that Sharon Layton was working on was a type of detergent that could be used for laundry as well as dishwashing and other types of cleaning that was non-toxic and plant based. This way, it could be produced on board, be renewable, and not too disruptive if it is accidentally made into the drinking water or food supply. No supply of this detergent had yet been delivered aboard. At this initial stage, everyone's laundry was just put into plastic bags and shipped home in empty Shrikes, at least for those actually living aboard full time. Also, Shrike food packets were being used as the only meals available, as the kitchenettes were not yet functional. This was no inconvenience to the seasoned spacefarers, though. The food wasn't half bad.

Assembly of all these living spaces took about a month. This seems pretty rapid in retrospect. But if you were aboard during that month, it seemed to take forever. Adam and Mattie, followed soon by Mangeet and Juni, took up occupancy in their own apartments as soon as they had been put together and tested. Each living area also had a large couch, a fair-sized table, and chairs similar to what was used on Shrikes. Both of these were equipped with restraints and could be used to watch the big screen televisions that were provided if the gravity conditions found you trying to float away. Pods were all to be equipped identically. No one was given more luxurious accommodations based on personal wealth or position in the company. They were plain but quite functional. In truth, the pods were more reminiscent of living quarters aboard a naval vessel than they were a cruise ship. On the other hand, each occupant of an apartment was free to bring whatever items they chose from home up to help decorate their rooms, size permitting. Everyone was also welcome to help to spruce up the ship as a whole. Both Thornes and Singhs, as the original two couples, continued to make occasional trips to the ground for purposes that were work related, but also to go shopping. The Thornes especially planned to stay in orbit a good deal more than they would stay in the mansion and were rather ambivalent about what to do with all of their Earth-bound possessions. Some of their favorite things were indeed brought aboard the station, but what would happen to the rest would be a process they would deal with later. Obviously, there was no room in orbit for any luxury sports cars. Maybe they could bring some art? They decided to publish rules for the crew, stating that all possessions brought up on Shrikes would need to be approved in advance by either Adam or Mattie. Otherwise, it was assumed that pandemonium may have resulted. But they did try to be at least a bit forgiving in this regard. Mattie decided to appoint herself as a sort of ship's purser and began to keep a log back in the bridge area of what was being brought aboard, just to keep things under some sort of control. Juni already was keeping a construction material manifest in the computer bay there. All materials and supplies were logged into the computer when brought aboard.

Bit by bit, and day by day, more habitation and apartment pods were assembled. Due to Manjeet's insistence and plain good planning, the ship's mass, or weight when G forces were applied, was distributed as equally as possible around the habitation wheel. This often necessitated that water-filled pods be interspersed between the apartments. Other spaces were allotted for agriculture and food production areas, Juni's entertainment space, gathering areas, and, of course, the extensive medical areas. All of these activities continued apace. Within another month, apartments began to become available for all the other members of the task force group. Not everyone came aboard at once, but they came all right. Generally, they arrived in the order they were needed, and when they did, they planned to stay aboard and make the ship their primary residence. Even the IRS was notified of the change of address. Sharon Layton was the next to move permanently aboard after the Thornes and Singhs, as food production is always the most vital concern for any crew. The widely assorted methods of this production needed as fast a start as possible. The big question with Sharon was that her time aboard any kind of orbiting spacecraft was so far quite limited. She was also single and was coming aboard alone. Adam hadn't really given this situation much thought prior to this, but he soon realized that everyone else working aloft so far had had some sort of support system. He and Mattie, as well as Manjeet and Juni, were married couples. Gregor and Manjeet were also teamed up as designers. Sharon did have a small team on the ground for food production, but she had not indicated that any of them were planning on coming aboard to assist. He would have to ask her how she felt about it when he went down next. Being the sole crew member responsible for feeding everyone for a very long time would be a lot to ask of her. One good development from her perspective was that the large rectangular arboretum structure was now being assembled at the back of the ship. This would be comprised of a few hundred panes of thick plexiglass in a huge aluminum framework, then filled with a pressurized atmosphere and kept heated to a friendly temperature for plant growth. The robots were busy around the clock putting it together. Sharon would be the czar of this new space, which would contain the largest volume of any compartment aboard. If nothing else, Sharon would certainly be busy. Meanwhile, the Shrikes arrived and departed, crews came and left again, and the big wheel was eventually set to its comfortable rate of rotation for the foreseeable future. This seemed likely to be a very long future indeed.

Part 3: Dining with Sharon

Sharon Layton had devoted all of her time since she was initially hired at Uplift devoted to the task assigned to her when she joined the Sustained Orbital Task Force. She had had no prior employment with the company and really had no aerospace related job history in her entire life. The job description was very simple sounding: "Design methods to feed a crew of twenty people, more or less, while totally isolated from Earth and in a closed system, with no hope of resupply." The rather unspoken corollary to this was that maybe, one day, the crew would choose to travel to some unspecified habitat with no resources for nutrition available on arriving there. Being a well-established theoretician in this field, she was the only person found who was a reasonable candidate for the job. She accepted mainly because she had never been employed at a place that would actually give her the resources needed to truly test her theories. This would not be the case at Uplift. Sharon was told up front that she had cart blanche in terms of a research budget, but also that she would absolutely be expected to produce tangible results. How could she refuse? She knew well in advance that she would generate all the results they could possibly expect, and then some. Her budget allowed her to hire two specialized assistants for this development phase of her soon to be enormous task. Her other more generic helpers were the Uplift staff people who would help figure out how to get her equipment and food sources into orbit and then install them once they got there. She was also allocated a large amount of warehouse space from which the cargos were loaded on the Shrikes. The first thing she decided to do was to prepare the twenty pods assigned for aquariums, which she considered a brilliant idea to have had built. These pods were divided equally between the freshwater fish and aquatic life, and seawater species. Replica rocks of minimal weight, as well as special mats that helped simulate a gravely seabed surface were brought up, then maneuvered into the tanks via the closeable trapdoor like hatches on each tank. These mats were manufactured so that they could be attached to the bottoms of the tanks, which retained their cylindrical shapes on the inside. Generally, these items were delivered during times when rotation of the big wheel had been halted and other large weighty cargo was being moved in. Once all the necessary materials were delivered to the small water tank anterooms, the big wheel was ordered to begin spinning to resume approximating one G. The anteroom air pressure was also cranked up a bit to help keep the water from backflowing when the access lids were opened. Thankfully, this actually worked to allow the aquariums to be accessed and the needed materials to be installed without water escaping from the tanks. Sharon and her few volunteers actually dove into the tanks themselves to arrange the habitats using air hoses which were available for each aquarium. For all this, she can be remembered as one of the first humans ever to go swimming while not on planet Earth. There were all sorts of firsts attained on this station build.

When the aquariums, aerators, pumps, and filters were completed, water in each of the tanks was heated to whatever temperatures were judged to be optimal for the species to be introduced. The tanks could even simulate tidal patterns, and lighting was in place to simulate what would be present in whatever depths were being simulated. Even the seasons of the year were reproduced. The next step was to begin introducing suitable plant life. Algae was always included. There were actually two pods devoted to nothing but algae since these organisms were quite competent in producing oxygen that could be used in quite a few onboard applications. Seaweed was placed next. Sharon had factored in which species of aquatic plants might also be used for human consumption. After being introduced into their various environments, a few weeks were allowed to let everything get established. When her satisfaction level was met, the Shrikes began to replicate the feat that was first accomplished by Adam Thorne on his very first command flight with Manjeet, Mattie, and Ana Koselka. After a somewhat lengthy hiatus, this idea was now being reproduced in a very big way. Huge supplies of fish would now begin to put into orbit in droves. This is not to say that they were packed in like sardines, but a few dozen of each of the chosen species were soon brought aboard. They were dispersed throughout various tanks so that their usual habitats were more naturally reproduced. Actually, prototype tanks had been built on Earth months ago to verify what combinations of fish populations seemed to work best together. Population densities, feeding protocols, and ways to best meet conditions to promote mating and reproduction of the chosen species were all documented and reproduced in orbit. This type of information was uploaded into the ship's computer bank for future reference. This was done in case Sharon wasn't around to supervise in the future. For whoever wound up doing the fish-related chores, maintenance of the tanks would take up the majority of their time and almost immediately became one of the primary shipboard tasks the crew would have to perform. The largest species to be included was Atlantic Salmon, and the smallest was shrimp. It was decided to forgo bringing up any Octopi from their home planet, but the smaller species of shrimp were not so lucky. Plankton was introduced as a food source too, as well as small fishlike anchovies (and sardines). Every species would have their preferred food sources available, and therefore, the humans would as well. No non-edible species were brought aboard. Getting all the various slippery little devils into the tanks was not particularly easy, but no aquatic animals were hurt in the making of this Space Station. Later on, especially at dinner time, not so much.

By the time this aquatic menagerie had been relocated to where few fish had gone before, the large arboretum had been completed. In addition to animal food sources (fish), vegetable ones were to be an even greater part of the ongoing dietary needs of the crew. This would necessitate that every member of the crew become a farmer as well as a fisherman. Humanity aboard the station would need to relearn skills that had been largely forgotten by the citizens of the modern world. Conveniently, this knowledge was also loaded into the ship's computers to rekindle that knowledge for the future. Edible plants were to be raised not just in the arboretum but throughout the service tunnels, in habitation pods, and anywhere else the environment would permit. In another first for human space flight, garden soil, with appropriate containers, was hauled into space in unprecedented amounts. Previous space-born gardeners had tended to go the hydroponic route so as to minimize the amount of dirt flying around the cabins. This was no longer especially the case for the Arboretum, which was expected to house, and soon did, various species of trees. A layer of air-permeable film was stretched over the soil troughs to keep down the dust. Water feed tubes would moisturize the soil from the bottom of the troughs. It would be a tidy arboretum if Sharon had anything to do with it. Nearly every sort of fruit tree, including figs, olives, citrus fruit, and even coconuts, was soon aboard and planted. Evergreens were also included, although none of the true monster species were actually planted for fear of them needing more space than could actually be provided over time. Seeds for all of these species were brought along though, just in case they were ever needed in the unspecified future. Planter trays and pots were soon popping up all over the ship and were fastened to various floors, bulkheads, and even ceilings, with their permeable membranes over the soil to hold it all in place in zero-G. Large amounts of fertilizers and hydroponic solutions were also stockpiled to last for several decades as needed. Storage for all this was created in the service corridors. Sharon and her assistants were planting seeds and seedlings as soon as all the pots, planters, and other infrastructure supports were in place. This included not only the various trees in the arboretum but also the larger plots dedicated to larger crops like corn, rice, wheat, and barley.

Sharon was also remarkably foresighted in including somewhat less nutritious foodstuffs in her inventory. Space was reserved in one of the service corridors for electric ovens, mixers, and kitchenware to allow for more conventional cooking and even baking. This would serve for preparing special meals and gatherings. There was plenty of stockpiling of sugar, flour, and things like chocolate chips, cocoa, and other flavorings. She also procured a machine for making milkshakes and a large supply of popcorn. Hopefully there would be periods with enough virtual gravity to allow for these things to be served without creating havoc. No less than five habitation pods were prepared as deep freezers for foodstuffs. As requested by other Task Force members, a large store of frozen turkeys, beef, pork, lamb, and even ice cream was placed aboard for occasional use, hopefully over a span of years. One commodity that would be less available than Sharon would have liked was dairy products. Those kinds of groceries simply didn't wear well. There would be large amounts of instant milk and a large quantity of frozen dairy products, but truly fresh milk simply wasn't really feasible. It would be possible to manufacture imitations from things like soy or almonds, but that would be the best anyone could do. In the case of future births aboard ship, there was a huge supply of powdered baby formula, but she instead would try to insist that breast feeding be the mainstay for infant nutrition if at all possible. A quite copious supply of canned goods was also ordered as yet another backup. Overall, Sharon had a plan to teach cooking skills to everyone on board to divide up the labor. She assumed that each family unit would do their own cooking. But eating was also a social skill that humans have always seemed to need to do in groups. This would be done occasionally in order to maintain some sort of community coalescence on the ship. She assumed group meals would continue to be held occasionally to maintain crew morale. Besides, who doesn't like attending a cooking class?

As a final measure, Sharon had always paid close attention to the fairly new field of artificial food synthesis. This science had been around for about 60 years and was constantly improving. Nearly any sort of meat analog product could be manufactured in vats with proteins and carbohydrates cloned from templates and manufactured with assistance from cultured bacteria. The commercial vats would not fit through the ship's hatchways, so she had three smaller ones specially made. If replacements were needed, they could be 3D printed aboard the printer that was also being installed in the same service corridor. Instructions for the procedures for these processes, as well as recipes, were loaded into the ship's computers. A good deal of Sharon's strategies developed on the ground were rehearsed and perfected starting soon as she was hired. Her overall goal of keeping the ship's crew not only well nourished, but also content with regards to the food supplied appeared to be being achieved decisively almost as soon as she came aboard. She consented to move aboard permanently about four months into the build, just shortly after making her first trip to the station. Happily, despite a good deal of pre-flight anxiety, she discovered that as long as the big wheel was turning to imitate about one G, she was able to accommodate the angular momentum in the way that most new sailors quickly adapted to the ship's motion. To her own surprise, she even found herself exhilarated when working in the larger weightless space of the arboretum. This was quite fortunate, as this area would generally be considered her primary domain. For her, it was a sort of office where she was most likely could be found when anyone was looking for her. Once settled in, she made only a few brief trips back to Earth for the entire time she was part of the crew. She soon became as indispensable to the ship's mission as any other crew member. No one was more surprised at this than she was, as she really didn't interact with the space-veteran engineer types much before she started working in orbit. Before very many days, she had sort of become the unofficial link to home, home being Earth. She was, in some ways, a mascot and brought plain and simple humanity with her to share with the other inhabitants of the station. She was forever exuberant, joyful, and as energetic as anybody on the ship. If humanity was to be exported to some unknowable destination, she certainly needed to be one of the people to export.

Of all the aspects of "permanent" occupation of the space environment, none apparently caused Adam Thorne more internal anxiety than the unquestioned need for on-board medical care. No crew could be expected to perform its duties competently for years on end if they believed nothing could be done for them if they became sick or injured. He was even forced to face the realization that this simple fact applied to him. The main problem for him was one that he all but admitted in the very first meeting of the Sustained Orbital Task Force. He simply was not comfortable with doctors. Perhaps he inherited this from his father, who really never saw a physician in his life as far as he could recall. Of course, Frank died accidentally when he was rather young. When he died, he had already expired before any doctor could arrive at the scene. This phenomenon of flight surgeons has always been a problem in the aviation industry. Anyone who has ever studied anything about the early days of spaceflight knows this to be true. Flight surgeons were the only guys who could take you off flight status with a single wave of their hands. All of those original Mercury flight candidates notoriously had to survive two weeks of hell while being evaluated (through every available orifice) at the Lovelace Medical Center. Ironically, this was located in Albuquerque, New Mexico and just a few miles north of the current Uplift Headquarters. None of those unfortunates who were evaluated there, most of whom became NASA astronauts at some time in the future, ever forgave NASA for having sent them there in the first place. They would tell their sordid tales about the experience for the rest of their lives, and nobody would recommend it be repeated for future flyers. Deke Slayton, who was grounded by the medics just prior to his scheduled Mercury flight for a heart arrhythmia, must have bitten his lip really hard every time he had to cope with the Fight Surgeons when he became NASA's Chief Astronaut, the all-powerful head of the

Astronaut Office. He wound up being one of the most important figures in the Apollo and Space Shuttle programs, a true legend. The irony of him dying later from the same chemical exposure that killed Frances Thorne made him Adam's main inspirational figure for everything he did at Uplift. Deke was the man.

It seems somewhat ironic in retrospect. But Adam quickly awakened to the fact that all of those ominous Flight Surgeons, who spent most of those 20th-century missions glued to monitors connected to the various probes and EKG pads the flight crews had to wear in orbit, were as vital to the missions as any of the various other personnel in Mission Control. Granted, the Docs eventually lightened up over time, but a physician was always required to be in the control room around the clock right up until the days of the first private space companies. Even Frances Thorne himself was forced by regulation to have an emergency room physician on call in Las Cruces for the company whenever flights were whisking off the planet, which was all the time. Several MDs were on the Uplift payroll at any given time. The company clinic on-site at Uplift always had flight surgeons on staff, and flight personnel were absolutely required to have annual physicals. Then, there was the association with the orbital space medical research station that Uplift had helped build and had supported strongly before it was sold. This station was now defunct for all intents and purposes. It was now destined to provide much of the personnel for Uplift's new, much larger orbital research facility. Dr. Peter Hosokawa, the always full-throttle surgeon, was now to be the spearhead of medical efforts on the station. Adam had reconciled himself to working with him and his now soon-to-be bride, Xing Xao Jia, an OBGYN with no experience in spaceflight. She was, however, highly motivated and energetic about giving this project a try. She maintained this Task Force assignment in addition to her private practice, which ironically enough was located at Lovelace Medical Center in Albuquerque.

She had volunteered for EVA (Extra Vehicular Activity) or spacewalking (to those few of you who don't know), training recently at NASA's submersion pools at Cape Canaveral. Peter had joined her, and they both earned passing certification grades without difficulty. Adam seemed to have not previously considered requiring this certificate for his Task Force members, apparently feeling that working outside the station would be the prerogative of the station's four robots. After being pestered about the topic by a variety of crew members though, it was impressed upon him that at some point in the future, anyone on the crew might be called on to leave the confines of the station. This might also be needed to be done from one of the Shrikes, for potentially unforeseen reasons. He and Mattie, as well as Manjeet, had already worked in EVA training, but this had mainly been instruction in the donning and operation of heavy spacesuits. So, he went ahead and scheduled himself and all the others who might permanently join the crew for full EVA training at NASA. NASA approved this, and classes would be held in increments over the next six months. This was one of those few reasons that Sharon Layton left the ship after moving in. Neither she nor any of the others expressed any trepidation about this requirement. Everyone was actually truly stoked about the idea, and everyone passed. Uplift, of course, had its own source for EVA suits, all made in the company's preferred navy-blue style. Two of these particular suits were ordered for each crew member, custom made. One was placed in the airlock in the hub of the small wheel, the second in one of the rim airlocks of the big wheel. For emergencies, a standby Shrike would always be present near the suits in both locations.

Outfitting the medical facilities was one of the last tasks performed during the station build. The vast amount of equipment and supplies required would have seriously gotten in the way of the earlier jobs, which had all manner of equipment and supplies of their own. It was thought best that those other, more vital requirements be finished first since they were more vital to those who were already living and working aboard the station. If there were medical issues for them, they could be quickly and easily transported down to New Mexico for treatment there. Aside from motion sickness, though, there were no real incidents aside from someone once bringing up a cold from Earth. During these early months, Peter's main task was the recruitment of medical personnel who might be a proper fit for duty on the station. General practice and simple medical qualifications for these jobs were not the only criteria needed to earn a berth aboard. Peter had been supremely cautious about who he would try to recruit. Many candidates might think taking this job would simply be a feather in their caps, but he had to make entirely certain they knew what they might be getting themselves into. He, therefore, developed a sort of "profile" that applicants had to fit in order to be considered. First, any type of flight-related experience was almost imperative. Second, there was the nearly unspoken qualification that real risk of personal harm or injury must be accepted. And thirdly, the candidate's family ties to Earth needed to be tenuous enough to allow long-term and potentially permanent separation. This had been a prime area of concern of Adam's from the start. As more and more of his plans entered the realm of reality, it became even more necessary to evaluate candidates before new members joined the crew.

Peter was slow to begin considering these factors when he first signed on himself. It gradually began to dawn on him that Adam must have given this some thought to himself when he selected the initial members of the task force. For example, Adam and Mattie were both orphans when they married, with no ties to life on Earth other than through the Uplift business. All this business was already now being done remotely from orbit. There were no family members to visit or be concerned about. Mangeet and Juni were in slightly different situations. Each of them had rather extended families. Adam was close enough to both families though, to know that they would happily support any decisions the two of them made regarding their careers. The families had immigrated to the United States from India and Japan fairly recently. Immigrant families, of all people, knew that sometimes you just had to pack up and go where life leads you. In both cases as well, both parents had worked all their lives in the Aerospace industry. This particular business existed solely to facilitate travel, any travel. If their children had chosen to help develop a trailblazing sort of travel, then so much the better. On reflection, Peter realized that similar situations existed for everyone in the Task Force. No one seemed to have any concrete reason not to abandon residence on the surface of the planet completely. People like Azaan and Gregor were former military men who had become disenchanted with both Russia and Tanzania for various political reasons. They were both unmarried with no family contacts, and both preferred the skies to the ground. They were also consummate professionals for whom their work was everything. Once they moved aboard, they each felt like they had gone to heaven without the dying part.

Iza Mazurkiewicz's family was originally from Poland but had to leave as World War II approached. Like innumerable other transplants from that most significant nation, they had no real desire to return if they found they could be of service elsewhere. Four generations later, Iza was born in Philadelphia, Pennsylvania. She was single, 28 years old, and far more interested in her career than her family's past. She had basically left her parents and the rest of her family behind when she went off to college in Albuquerque, where she was a later addition to the Number Nines. She had had very little contact with her family ever since. She never did reveal why, but things like family alcoholism and some parental abusive behavior had been hinted at in her interview with Adam. Iza had no regrets about the separation. She had come aboard as soon as there was space available for her on the station. She had her pod assigned to her and jumped in enthusiastically decorating it. She had already worked feverishly from the ground, making sure her four precious robots were performing to everyone's high expectations. She was glad to be able to cut to the chase and watch over every move they made at close range. She was always too hyped up and fascinated with life aboard the station to bother not loving every minute of it. She used her surplus of energy to help Mary Dupree supervise the placement of the smaller onboard 3D printer in one of the service corridors. A much larger one was being built to be mounted in the engine spaces beyond the radiation barrier for Sparky to use. Much of what was planned for back there was a secret at the point she came aboard, even to her. But her qualifications and aptitudes made her the best surrogate for Mary Dupree, who was still needed most on the ground at that point. Mary was, in fact, the last of the original task force members to come aboard. Innumerable updates and changes in the manufacturing plans were happening basically every day, and Mary, with her team of manufacturing designers, was always able to take up the challenge. They always seemed to deliver on time. When it was finally determined that any further manufacturing needs could be met with the printing capabilities already aboard, she finally flew up permanently.

Mary, of all the crew, actually had made the fewest Shrike flights. She did get her EVA training before the move, at least. She, of all the crew, seemed the most introverted and introspective. If she had any real problems getting used to zero-G, she kept it to herself. It was eventually discovered that she was basically estranged from her family, who had all lived in Detroit, where she was born herself. She showed only pure professionalism at all times and was really rather surly. She was also only 25 and was sturdily built. Her hair was currently purple, and she worried about how she would be able to continue dying it in zero-G conditions. She was suspected of not entirely disliking Gregor Levinski despite his Russian heritage. It was unknown if Gregor returned these feelings, at least at this point. Her parents were both elderly and lived in an assisted living facility in Albuquerque. Her brother had the power of attorney for both of them. They both suffered from dementia and had no idea that they even had a daughter named Mary. Her brother was also an actual attorney who had no interest in whether his sister lived on the planet or not. He certainly didn't include her in any decision-making regarding their parents. The two were definitely not close, to say the least. Family ties were minimal in her case as well.

But we digress rather badly. We had actually been discussing Peter Hosokawa and the medical team. By the time the earlier cast of characters were tenuously established as the engineering portion of the prospective crew, Peter was fully concentrating on finalizing who would comprise his medical staff. His challenge had always been to find the minimum number of personnel that could cover any and all conceivable medical needs that might arise aboard the station. This group would hopefully sign on for duty to cover the entire lifespans of a crew of twenty, more or less, plus any additional patients that might be born aboard the station in the nebulous future. The services to be provided ought to include staffing a tiny hospital, surgical suite, dentist's office, and family practice clinic, and that would be the bare minimum. If they could also staff an oncologist's office, orthopedist's office, physical therapy facility, and optometrist, that wouldn't hurt either. Peter couldn't help but envision Adam's recitation about the early ocean-going physicians at the dawn of seafaring explorers who had been called upon to do it all. On reflection, he decided he didn't want to emulate these poor fellows in the slightest. He first called upon the professionals he had worked with for a few years on the orbital zero-G clinic he had left around about two years ago now. He also turned to the unofficial criteria that Adam had used when selecting his people. Simply put, candidates should be drawn to the idea of working in space and also not have especially close ties to anyone they might be leaving behind. At first, it seemed like no such people existed, no matter where he looked. And if they did, for the first few weeks it seemed as though he had no prayer of finding any of them. Nevertheless, since he would certainly never find them if he didn't look, he kept an open mind and kept on looking. He eventually figured it out.

The best place to look was the military. He couldn't believe he hadn't thought of it before. This line of inquiry had its share of hazards, though, since involving any portion of the US government in this project could irreparably complicate everything the task force was trying to accomplish. This was an obvious fact. The thing to do was look for people who would be willing to leave their respective services behind. It would be even better if they had done so already. The Air and Space Forces were the obvious first candidates. Anyone who had already flown with NASA or other space-faring groups got the first look-over. He had to do his searching in a very inconspicuous fashion, but to his surprise, he soon had a rather thick file of possible candidates. Copying NASA in the early days, Peter's favorites soon received letters or e-mails asking if they might like to come and interview with Uplift Aeronautics. If so, could they please fill out and return a questionnaire that subtly asked questions about suitability using Adam's suggested criteria? He was pleased to find about twenty candidates total who could fill the specialty positions he wanted to be staffed. The first of these was a Pharmacologist, or at least a pharmacist, to handle the medication and medication therapy side of things. As an afterthought, he had also sent applications to several drug companies. To build a supply of these medications, he sent letters to drug company sales representatives to say that the new Uplift Orbiting Research Facility was seeking to stock a large pharmacy for, well, "research purposes." Companies that donated medications could then crow about their donations for advertising purposes. He then began interviewing his candidates and awaited responses from potential sponsors. On a whim, he also started asking for handouts from medical equipment companies just to see what he might get donated. Within about three months, he had found a small group of candidates who felt as though they would be willing to try staffing the station, at least for a trial period. There were more people who qualified than he originally imagined he would find, but Adam had told him to hire as he saw fit. There would be room aboard the ship for at least ten of them, counting him and Xao Jia. Here's who he picked:

Sandra Annison MD, a General Practitioner from Chicago, Illinois. She was also a former registered nurse and an Air Force veteran. She had been a former NASA astronaut who conducted research on the ISS for a total of 233 days. She was single, and somehow only 32 years old. Her experience included knowledge of medical laboratory science and procedures, having already done blood and serum testing aboard the space station and interpreting results. She was also already received EVA training but had never been called upon to use it.

Victoria Zasco MD, an Oral Surgeon from the US Air Force. She was 28, a bit of a prodigy, but had worked in field hospitals near several war zones and had combat training as well. She was a native of Tuscon Arizona and was also a bit of a fitness expert. Educated at Norte Dame as an ROTC cadet, she attained the rank of major before retiring to private practice. Also a qualified dentist, she even had experience with 3D printing of dentures and implants. She apparently also could use computer software to construct orthodontic braces for pediatric and adult patients. She felt she could also assist with surgery to the face and neck with computer assistance. Peter, as a surgeon, had worked with this type of assistance a time or two, so that helped finalize her selection. Victoria also was single, and her parents in Tuscon were well accustomed to her duties calling her away. "They'll support me whatever I'm involved in," she said in her interview, "as long as I can get a message to them once in a while." He came right out and told her that travel to deep space might be an objective for this task force he was assembling. He said that communication strategies were being worked on, but no promises could yet be made. She indicated that it was satisfactory. Victoria had never been in space but seemed quite enthusiastic.

Kevin St. Patrick MD, a 27-year-old Pharmacologist. He was recruited from Johns Hopkins, where he was the co-director of the main pharmacy. A captain in the Space Force, he would accept this Uplift position as a detached assignment. He would still be considered as active duty even while aboard the Uplift station. One of the rather scandalous parts of his inclusion was that Kevin had no idea at the outset that anyone was considering making this station into an actual space vehicle until fairly late in the program. No one knew how he would react, and so everyone went to great lengths to hide this information from him until Adam decided the time was right. He was ultimately selected anyway because his father had recently died of cancer, and his mother was completely estranged from him. In high school, he was nominated to attend the Space Force Academy and quickly accepted. He received a degree in pharmacology from Johns Hopkins and was quickly assigned there. He was a First Lieutenant when Dr. Hosokawa stumbled on his name and then hired him. He happened to be promoted to Captain just after this. At his interview, Adam saw a sort of kindred spirit in Kevin since his situation was nearly identical regarding the death of his father. Kevin's chemistry with the group was immediately obvious. He had joined the Space Force looking for adventure, but he wasn't really finding it with his current assignment. It seemed to him that he would never leave the confines of stateside hospitals and clinics, no matter how long he stayed in the service. He was from Washington, DC and hadn't seen his father from the time he started at the Space Force Academy at the age of 18. He died three years ago. Kevin had no experience off planet but had been obsessed with the idea of spaceflight since his time at the Academy. He was exceptionally outgoing and buoyant, plus he really knew his medications. He also knew how to manufacture most of them. Peter consulted extensively with him when designing the medication manifest for the ship.

Tariq al Abdullah, was a 30 year old man from Morocco. He was 6 feet 2 inches tall and 195 pounds in weight. Tariq was a neurosurgeon educated and living in the United States and working at Vandenberg Air Force Base. A naturalized US citizen, he was also a retired Air Force captain who had conducted a bit of research aboard a Space X orbital mission. Single, and basically exiled from Morocco, he felt no ties to anything except the United States, and he had never been stationed abroad once he moved to the US. Peter felt that a neurosurgeon would be a vital addition to the crew, as neither he nor anyone else had experience in that field. Neurosurgeons also were involved with things like motor rehabilitation and physical therapy. Peter also felt that Tariq could fill the need for an Ophthalmic Surgeon and vision care provider. Audiology and Ear, Nose, and Throat problems could easily be handled between them. It felt like all the pieces were there, and so Tariq was accepted.

Peter decided that these additions to the crew would be adequate, but only adequate for the mission to proceed. Salary negotiations were completed in Uplift's Human Resources department, as were health and life insurance plans. The medical people seemed more interested in these things than the rest of the crews, but they were accommodated to their satisfaction. In most cases, due to the lifestyles of the selectees, beneficiaries were rather hard to specify. Those who had never been to space were sent to NASA for EVA training before taking their first visits to the orbital station. They would then participate in outfitting their various pods and getting settled in their living spaces. By November 2086, the medical staff and equipment had reached a point where they were ready to come aboard their new home, supposedly on a permanent basis. How long they would actually choose to stay was anybody's guess. Peter Hosokawa and Xiang Xao Jia simplified living arrangements by electing to share a single apartment. They were not yet married, but were liberal enough to live together with their engagement already arranged. They chose to make English their primary language on board, although they had each learned a good deal of their future spouse's languages of Japanese and Mandarin Chinese. All of the other medical staff were assigned their own apartments. The work of lifting all of their medical and pharmaceutical equipment and supplies was the last major task to accomplish before making the station "operational," at least as operational as the orbital platform was officially meant to be. This was every bit as herculean a task as any of the other phases of the build. But medical professionals are nothing if not used to extremely hard work.

Peter at least had been successful in persuading major medical companies to donate small fortunes worth of equipment and supplies. These deals tended to hinge on these companies getting permission to come up to the station to make commercials for themselves. The single largest donation was from the company that would be providing the MRI machine, x-ray equipment, and the fluoroscope used for things like cardiac stent placement and cerebral and peripheral circulation assessments. These machines together would have cost in the tens of millions of dollars if they hadn't been donated. Recent advances had reduced the weight of these devices to the point that the pod they were placed in could easily be balanced with a water pod placed opposite it on the big ring. This was a requirement strictly demanded by Manjeet when the ring was initially designed. These machines required a fair amount of power and support equipment to operate. Things like radiation screens for the radiology tech, coolants, and computer support were needed in this pod. Restraints were needed on many of the devices in case they were being used in zero-G. Like the habitation pods, all of the equipment needed to be rotatable within the pods for the various thrust scenarios as well. The equipment needed to be capable of being broken down into smaller sizes than the assembled final products in order to fit them through the corridors and hatches and get them to their final destinations. Poor Kevin St. Patrick had quite a logistical challenge in fitting his pharmacy medications and supplies into a single pod, so he asked for and got two. There was also a hood installed where various ingredients could be manufactured and combined if replacement drugs had to be made, or if anything was needed that wasn't already aboard. Storage space for refrigerated substances, such as vaccines or blood supplies, took up valuable real estate in the pod as well.

During this period, a blood drive was held at Uplift to meet the need for an onboard supply for the crew. O negative, the universal donor, was the type of choice to be collected. The crew had a variety of types to cover. Besides O negative, any other matching types were accepted. As a point of curiosity, Adam Thorne's blood type happened to be AB negative, the rarest type in the United States. Strange that such an unusual individual had this particular blood type. He could accept O negative since it was devoid of either A, B, or positive rH. Then there was a huge supply of general oral medications needed. These were stored in rather plain shelves and cabinets with steel bottoms located in one of pharmacy pods. Small magnets were placed in the bottoms of the variously sized bottles by the manufacturers to hold them in place. There were numerous cubby holes and doors on the lockers with lists of medications and maps of what was inside. Since the pod was tubular in shape, lockers were placed all around the circumference of the tube. Kevin could float around anywhere in zero-G to retrieve what he needed or actually walk on lockers when he had weight and use them as floors when needed. He could also rotate the inside of the pod to whatever orientation was best to help him access what he was looking for. This helped if he needed to use the hood area or keep liquid containing bottles with the lids facing "upwards." It was a constant challenge for him to use the space to its best effect, more so for him than any other member of the crew.

A supply of IV solutions were stockpiled, as were medical gases and anesthesia to be used if surgical procedures demanded them. IVs had been infused with pumps for many decades, so no gravity was required for their use. The surgical suite had one permanent rotating table, and another fold down stretcher. It was hoped that no more than one emergent type of surgery was to be performed at any one time, since only one fully qualified surgical team would be aboard, but one never knew what might come up. The robotics company that had donated the four exterior construction bots also chipped in two robotic surgeons to be used for such times as no humans were available. They could even do very precise microsurgery if the need ever arose. There were mounting stations on the floor around the surgical table if the bots were ever called upon for service. For the time being, they were stored in the same service corridor as the 3D printer. The single surgical pod could also be used for obstetrical procedures when needed, and most of the crew realized that it would definitely be necessary if there was a need someday. Surgical instruments, lighting, and that most necessary suction machine with a very large collection tank had been procured and was quickly installed. Fluid loss was to be expected when giving birth or with various types of surgery. If all of this occurred in zero-G conditions, there had better be such a suction machine nearby. Again, the entire suite was rotatable to accommodate whatever G-conditions might apply at the time. It was also enormously helpful that an Earth-like atmosphere was to be recreated on board. Electrical cauterization machines to control bleeding were an absolute necessity, and these would cause explosions if used in a pure oxygen environment. Finally, actual magnetic footwear was created for those performing and assisting with surgeries. A steel floor was placed below the surgical table. In this way, those hovering over the patient would not be actually hovering. They could stand with their feet securely on the floor, although moving their feet once they were planted down would be a bit challenging. Peter Hosokawa was very proud of the operating room he had created, and justifiably so. He just hoped he would not have to use it too often. He hoped even more that he would never have to lie on that table himself. He made a note to ask Xao Xiang to review surgeries like C-sections and other gynecologic surgeries with him in case she was the one on the table someday. In that case, he would be replacing her as surgeon if she were the patient.

The next pod dedicated to active medical practice was reserved for Dentistry and Eye care. Here were placed the traditional Dentist's chair, bolted to the floor, with two stools for the dentist and his assistant as well. There were sturdy restraints in place to hold them securely in those chairs. An assistant for Victoria had yet to be appointed from the crew, although Kevin did have a practical knowledge of things like use of Novocain and so forth. Victoria was actively trying to get him to volunteer. The usual dental equipment with drills and instruments were arranged as in any dental clinic. You could spit in a cup connected to suction, so there was a good chance of the yucky stuff not floating away in zero-G conditions. The equipment also for eye exams and manufacture of both eyeglasses and contact lenses were also in this room, as well equipment to make dentures, crowns, and bridges if needed. Dr. Zasco asked all crew members to schedule their first on-board dental exams as soon as possible, hopefully within the first two months aboard after she came up. She could then enter a database for everyone using the computer terminal in her office pod. Getting everyone to comply with her request proved to be challenging. On Earth, she was almost exclusively an Oral surgeon, and patients were sent to her due to pressing need only. General Dentists, though, nearly always have to bother their patients incessantly to convince them to come in. This apparently was the case everywhere in the universe. Tariq al Abdullah similarly requested all crew members to schedule eye exams to help him develop charts as a baseline for everyone's future care. This was not too challenging since none of the crew had as yet ever developed problems with their currently perfect vision.

Finally, a pod was specifically assigned to Dr. Sandra Annison, the General Practitioner of the group. She would be the one who would address the crew's day-to-day problems and would know everyone's most intimate health secrets. The computer housing each crew member's health histories and test results, their medical charts really, would be located in this pod. The interior resembled a typical doctor's office examination room. There were supplies for general exams placed in secure drawers and lockers. Scales for adults as well as infants were placed there but could only be used when any sort of gravitational type forces were in play. No one had yet invented a zero-

G scale. A small supply of everyday medications was stored there in a closet sized locker. Things like wound care supplies, thermometers, and sterile wipes and gloves were all there as well, just as in any Earth-bound doctor's office. She had no nurses or assistants and was pretty much on her own. She, like all the other medical staff, knew that most of their time would not be spent practicing medicine. All the medics would be mostly engaged in everyday crew activities like cooking, cleaning, tending to plant maintenance, cleaning aquariums, and standing watch. They would be assigned to jobs daily by First Mate Mattie, exactly like the rest of the crew. But all of them would generally start each day at their own professional pods reviewing the status of everyone's health. There, they performed daily inventories, checked their equipment, and saw patients as needed. This happened more often than one might expect, since it was imperative to keep everyone as fit and well as possible in order to perform their duties. There were various pieces of monitoring equipment that Sandra kept in her pod, and she was the one charged with maintaining them. These included things like oxygen monitors, glucometers, and testing monitors that could analyze blood counts and chemistry from only a fingerstick. It was Sandra who insisted that a mammogram scanner be brought up to the station, despite the weight and expense incurred. Other lab equipment was located both in the OR and the pharmacy. Several portable oxygen concentrators were available, although if anyone wound up needing supplemental oxygen long term, then the atmosphere in that person's pod could be increased selectively. Theoretically, if more exotic testing equipment were needed, Mary could make it in short order with the 3D printer. All in all, it was a medical facility comparable to or exceeding anything you might find on a top-of-the-line US naval vessel in the 2080s. The medical pod area was apparently never referred to as"sick bay" by the crew. Perhaps they didn't like the Star Trek references. It probably just had too much of a military sound to it that just didn't seem to fit. Whatever anyone called it, the crew hoped these ample resources would not be needed too often. When it was completed, no one could say how healthy this new ship would be, or for how long.

A significant part of Adam's strategy for attempting to have his people commit to living an entire lifetime in an enclosed space for the rest of their lives while totally isolated from any possible relief from this unusual situation was to try to provide significant distractions for the crew. He knew from his studies of history that a multitude of seafaring people, mostly men, had also spent their lives aboard wooden ships with little or no contact with the outside world. Later on, those with professional seagoing careers eventually reached a stage where they despised any sort of life away from their beloved ships. Making landfall became merely a necessary evil, though walking on land would be a sporadic and very temporary occurrence. But at least these men had the comfort to know that they could almost always count on things like air to breath and food to eat. On the other hand, the ocean could drown a whole ship's crew in a few swift minutes. Storms could smash a ship into coastal rocks, and disease or combat often would decimate hundreds in weeks or even minutes. It fascinated Adam that the human need for adventure, conquest or love of country ensured that sailing vessels were always fully crewed from the time that ocean-going ships were invented. Even in the days of the early explorers, when no one had the slightest idea what awaited them just beyond the next horizon, humans just grinned broadly and sailed on. Adam knew he would need to tap into this inborn human characteristic of curiosity, as well as selflessness, to have any chance of turning his ideas for his nearly completed research facility into something more. He found himself amazed that none of his selected companions ever balked or questioned little clues he sometimes mentioned that hinted at his aspirations. True, initially these hints were very subtle, but it seemed to him that perhaps his task force might actually share his point of view. So far, everyone winked and agreed that all they were doing was research into how one might actually leave the Earth behind without intending to return any time soon.

So, if one might leave their home port, what small portions of human existence would it be advisable to take along on the trip? Some of these issues had already been brought up in task force meetings. As an example, Juni had already begun to build and was completing a pair of pods with projection screens, flat-screen TVs really, that could project images of various environments all around the occupants. One could have a simulated picnic in the woods, a trip down a river in a raft, flying over the Grand Canyon or the Alps, a concert at Carnegie Hall, or whatever else you might wish to experience. Sound systems could augment these environments as well. One could even change the ambient temperature in these pods if desired unless you wanted something extreme like a volcanic eruption or something. Another absolute must for crew health was to bring exercise equipment aboard. This sort of thing had been done as early as the Skylab project, when astronauts had a large interior room made from the inside of a hollow Saturn V booster shell. The crew could run around the periphery in a circle, doing handstands and flips as they ran. Little exercise bikes and elastic bands were also on board. On this newly minted station, the two auxiliary rings, or "service corridors," could also serve as running tracks that anyone could use whether the rings were turning or not. Though partially filled with equipment, vents, pumps and all manner of other occasionally used items, these were arranged so that straight walking paths were available regardless of where the acceleration forces were acting at any stage of the flight. This principle was followed in all habitable areas, and great lengths were taken to make it so. The large arboretum area, now mostly completed, could also be used as a place for relaxation and even solitude. Even though one would need to float freely in the chamber, at least initially, it could feel something like swimming or scuba diving. It was realized quickly that you would need to secure yourself to a wall with some sort of line or tether when doing this kind of thing, since if you ran out of momentum somewhere in the center of the large room, you might just hover somewhere in the center with nothing to grab on to. If you didn't run out of momentum, there would be nothing to keep you from being plastered against the plexiglass. Sharon also requested and got things like a projection device to put images from Earth landscapes up on one of the plexiglass walls. On request, you could even ask for a movie to be shown, drive-in style, for movie nights that were often attended by the entire crew.

Also, not one, but two pods were dedicated as gymnasium space. The first was a makeshift weight room, complete with the sorts of zero-G equipment like the ones made for previous orbital vehicles. These included resistance-based machines for leg and upper body exercises and stationary bicycles. These could also easily be used in simulated gravity. Then, there were areas holding traditional weights, benches, squat racks and other assorted machines that could be used when in various gravity simulation conditions. Like all the other pods, the contents could be rotated internally to fit the acceleration situation. There was unfortunately no adequate space available for team sport activities like basketball or soccer. But who was to say something couldn't be figured out later? The computer system was loaded full of recordings of Earth-bound sports that you could watch in your pod, or in group areas if there was interest. One such area was the old bar and lounge located in the small wheel, from back when it was still a hotel. Though only a fraction of a G when spinning, a large folding meeting table was ordered placed by Mangeet that would allow the entire crew to assemble there for meetings or social gatherings. The usual strategy of harnesses for the chairs and Velcro on the table was used here. When the station assembly was essentially complete, task force meetings were increased to weekly, as opposed to the earlier monthly meetings. Movies and other presentations could be shown on a large LED screen. Anyone could reserve time for their own use of this space as much as they wished. The only drawback was the relative difficulty of getting to the meeting room and the changing acceleration forces you experienced making the trip to get there.

Getting back on topic, the second recreation pod, next in line from the gym pod, was dedicated to hobbies. One main focus was music. As advertised in planning meetings, a variety of instruments were located in lockers. Two guitars, one rhythm, one electric, an electric bass, a keyboard, a clarinet, a trumpet, and a slide trombone were chosen as the instruments of choice. There was also a synthesizer that could emulate a variety of orchestral instruments, as well as an entire orchestra. A bit later in the build, a violin and banjo were requested and brought aboard. Oddly, no crew members ever had had the time in their Earthbound careers to indulge in learning to play an instrument, but no one asserted that they would never attempt to do so. There was yet another flat-screen television provided in this "Music Room." The computer could provide and play any of the music loaded into the servers. This included pretty much any musical piece that had ever been recorded up until 2086. There were also music lessons available to watch as taught by eminent music teachers from all over the world. There was one set of drums provided in the music room. A rock and roll set had been bolted to the floor in one corner, but if the gravity situation wasn't just right, hitting the drums with your drumsticks might send you flying off in bizarre directions. Although his room could also be rotated like the rest of the pods, perfect gravitational forces to accommodate drummers were not usually possible to create. This problem would require more research if anyone ever cared to conduct it. Mary Dupree reported that she had seen to it that appropriate materials and programs were in the computer protocols in case anyone wanted to 3D print any other instruments. The only thing she didn't think could be made in this way was a set of bagpipes, but no one had expressed any interest in doing this. No tubas were ever requested, either.

These were the only measures that had been requested by the time the interior features of the Orbital Laboratory were essentially completed. All the task force members who had participated in the design process had by that time volunteered to live full time aboard in their own habitation pods. No one as yet had been asked to make any permanent commitments to live on the station permanently and could still conceivably quit and go back to work on the ground any time they chose. Two members who had not yet made the station home were Azaan Nazari and Awamila Hawat, as they had not yet reached a point in their respective fields of responsibility to allow them to come aboard. Once Awamila's astrophysics lab area was completed on the "underside" of the main frame, just below the arboretum, she signaled she would come up to finish fitting it out. A large telescope, as powerful as the legendary Hubble, but requiring a heated area with an atmosphere, had been designed and built by a company which mass produced home telescopes for sale to the public. Awamila found it to be capable enough for the job required, and it was donated to the station project in exchange for advertising rights. It was fairly large, but could be disassembled, fit aboard a Shrike, then reassembled once it arrived in her lab. The remainder of her laboratory equipment was either purchased, donated or already owned by Uplift. This portion of the ship's equipage was completed in early November of 2086. With this accomplished, the habitable portions of the station build were now completed. Mila filed a change of address with the US postal service. All her mail would henceforth technically be airmail.

Chapter 13
A Year in Orbit

*From compiled recorded orbital meeting minutes and revealed
events after the fact..*

With Awamila aboard and now installed in her own apartment, the only crew member still missing was Azaan Nazari. He had been spending the past several months down at Uplift grappling with the most challenging of all the assignments, with all its accompanying problems, of the entire project. His unenviable job was to determine whether any sort of propulsion system could take this new station anywhere that was not Earth orbit. If in fact he could, he would also need to try his hardest to make sure that not too many human generations passed before the ship reached its designated destination. To help him with this task, Adam had ordered a secret underground laboratory to be built for Azaan's personal use. Great lengths were taken to conceal Azaan's work from the general public. Not even Nathaniel Floatingfeather, the de facto manager of Uplift Ground Operations, caught wind of this secret lab until it was too late to do anything about it. He surely would have quashed this work if he had found out sooner. On November 15th, 2086, Azaan rocketed up to the station to make a status report at a task force meeting.

Adam went through the usual role call for department updates. Mattie spoke first, presenting a new duty log system where all crew members were assigned individual tasks for the month on a calendar available on everyone's cell phones or on the general computer access laptops issued to all members of the crew. Everyone was required to bring these laptops to each meeting. Daily individual duties usually began at everyone's usual workstations. This was especially true for the medical people, since their workstations were all in sick bay. One crew member would always be on duty in the "bridge" or main control room area, and someone would be rotated there every eight hours around the clock. All control functions could be placed on automatic, so that whoever had the night watch could sleep in the hammock at the back of the room for most of the shift. This night commander would be awakened by the computer if any alarm should happen during the shift. US Mountain Standard Time, used on the ground at Uplift, would remain in effect on the station in perpetuity no matter where it happened to be. The night shift was 9 p.m. to 7 a.m. After a morning inspection of each member's primary assignment station, the work schedule would divide everyone into groups to attend to such things as crop and food production, food preparation, and inspection of all the ship's internal equipment. Status reports from each group would be entered in the ship's log via the laptop of the assigned group's supervisor. Any unusual situations that were discovered, like a piece of equipment that needed overhaul or redesign, would then have personnel assigned to resolve it by Mattie. She would act as a sort of "first mate" on the station, with Adam as the presumptive captain. Peter was the currently assigned head of the Medical Department. Manjeet managed the Engineers, with Iza, Mary, and Gregor under him. Sharon had food production and coordinated leisure activities, Awamila physics and astronomy, and Azaan propulsion. It was obviously vital that everyone would need to be cross-trained as much as possible. It was then expected that all would eventually assist in the full gamut of vital on-board activities. The medics would need to be the most flexible, since they would only be actively practicing their professions for a small fraction of their total time aboard. Fortunately, all of them were highly intelligent people, with flexibility being a major requirement in the medical profession anyway. So far, the crew was meshing extremely well.

Department by department, Adam went through the status reports posted so far by the entire crew. The fish appeared content in their aquariums, and protocols for their care and feeding were being developed and improved. Sharon reported that efforts to plant and cultivate the myriad of vegetable seeds and sprouts was ongoing, and that she would request as much help as Mattie could schedule for her. For now, the food supplies brought up from the ground would be able to feed them for a further three months before any renewable food sources needed to be utilized. Manjeet reported that power management, atmospheric maintenance, heating, cooling, and overall ship performance parameters were performing well above expectations. Awamila was calibrating the new telescope and working on the implementation of other monitoring and research equipment she had brought aboard. Kevin was cataloging his inventory of medications and other supplies and would submit an invoice to Mattie regarding any future needs for delivery as soon as possible. All medical equipment checks done so far had been successful. The physicians reminded everyone of the need to schedule their required exams as soon as possible. Radar and communications were operating without difficulty, but Juni reiterated the need to further develop possible methods of deep-space communication. This was all so far merely theoretical. All this was duly noted. Iza reported that continuing final construction tasks were being accomplished as expected with her four bots. The crew was always fascinated to hear about the mechanical marvels and a sort of mythos was being built around them. Joking about the bots seemed to help with crew morale for some reason. Heroic and folkloric stories were being told about them amongst the crew on an almost daily basis. People would ask, "Did you hear what Sinbad did yesterday?" and then answer the question with some tall tale or other.

After this necessarily compact summary of mission status issues was completed, the medical people were dismissed. This was an unusual thing for Adam to do, and Peter called him out about it. His excuse to them was that the rest of the meeting would address only engineering issues and would only bore them anyhow. Peter did make the objection that the medical crew members were supposed to be considered full partners in all crew decisions. Adam replied that some of the issues to be discussed could be misinterpreted as policy, and it would be best to have as few people involved as possible to prevent gossip or anxiety about "what we're trying to accomplish up here." "Besides," he said, "I don't like the idea of having the rest of the station left alone while we're in here being a bunch of geeks. Believe me, if and when we do anything that might impact any of you in any way, you won't be asked to leave." Peter shrugged but grudgingly led his people out of the room. Unfortunately, this small incident did create a bit of distrust between the engineers and the non-engineers that Adam would eventually have to smooth over later. It would reinforce the fact that it was never a good idea to stratify such a small crew. Actually, the underlying reason the medics were asked to leave was that Kevin St. Patrick was still an active-duty Space Force officer. What they were about to discuss would be almost definitely illegal if it were ever implemented. If nothing else, Kevin would absolutely have felt obligated to report what was discussed next to his immediate superiors if he had overheard what was said.

"Well," Adam resumed when the others had left. "How goes the propulsion side of things, Azaan?"

"Not quite in the way I envisioned," he answered. "But things could improve drastically much sooner than I thought."

"This should be interesting," Adam said. "Please go ahead."

"First of all, I see no reason to think we can't proceed right away with building and installing the ion drive engine we've been talking about. I've sequestered a very healthy supply of Argon gas and have even tested a prototype at one-quarter gage size. The computers say it will work, and so do I."

"OK good," was the general sentiment.

"That could get us to the gas giants faster than anyone has managed so far," Adam said.

"But that's not all we talked about, was it?" He knew most of those present had not heard much about any other projects, only a few speculative murmurs.

"No, it wasn't," Azaan continued. "As to the antimatter engine idea (muffled murmuring is heard on the taped record), I found it was completely non-feasible to begin collecting any antiprotons and storing them in our underground lab. (Questioning gasps this time). For starters, there is no way that any activity of that kind would go undiscovered. Secondly, any accidents with the storage equipment would have destroyed the whole Uplift facility and possibly killed everyone we know and love, and even more that we don't."

"So, there's no hope of building the kind of engine you envision?" Adam said disappointedly.

"No, I'm not saying that at all. I'm saying we couldn't have developed it on the ground. The promising part is that I realized, with Awamila's help, that if any antiprotons are indeed present in the Van Allen belts, it would be concentrated in pockets created by the Earth's magnetic field lines. We would read between the lines if you will. Now no one has done any recent mapping of the field lines for quite a few years. The fields are always changing, for starters. For us orbital dwellers and all other space travelers, we just know to stay the hell out of the Van Allen belt if we want to have children and live long lives. It certainly would be a bad idea to just cruise around them in a Shrike, hoping to catch antiprotons in some kind of magnetic trap."

"I see your point," Mattie blurted out in a very Mattie-like way. "So, are there any ideas about this that aren't bad?"

"To answer that, let's consider the storage problem," Azaan answered. "If we can actually gather up the necessary ten grams of antiprotons and can't risk bringing them back to Earth, then what can we do? "

"Bring them here!" Awamila screeched. "That must be why you haven't begun to collect them already! Listen everyone! What Azaan and I have actually accomplished during that time on the ground was to design and build a few dozen small magnetic containment vessels that can maneuver themselves around independently to collect antiprotons. They can detect and search for them on their own. When they collect a quarter of a gram, they can maneuver out of the radiation zone and alert us. Then we can go collect them with a Shrike with a lead-lined containment chamber and bring them back!"

"To a large containment tank that will be located somewhere on board this station!" Adam finished the thought. "But where on board?"

"That's next," Azaan continued. "Manjeet and I have been corresponding. We've made plans for the structural designs of the trusses and components for everything that we'll need to build aft of the Arboretum and observatory to create our engines. First, we'll need a pretty large shield structure to deflect any radiation away from our living spaces. Then we'll strand Sparky back there to do the construction work. There will need to be a huge tank for liquid hydrogen to provide cooling for the reaction chamber. There will also be a large array of radiant coolers to reliquefy the hydrogen once it's run by the superhot conducting shafts and engine bell. Then there are the massive magnet arrays generating the containment fields for the antiprotons. We'll generate a plasma of hydrogen ions and shoot them down the main conduction shaft, and then inject an antiproton, just one at a time, to combine with the proton stream. Not too many hydrogen ions will need to be shot downstream, but when a proton and antiproton combine, they'll interact out the back end at hypersonic speed. That will create a massive explosion that's already moving incredibly fast. Then voila! Continuously accelerating thrust that can last for years on end. When we get to the proper point in our trajectory, we wheel around 180 degrees and start slowing down. We've run our designs through the Uplift computers, and it seems quite feasible. We'll place a huge 3D printer at the back of the ship to build the engines. We should use multiple layers of synthesized Borophene to handle the heat loads and stresses. There will be networks of tubes built into the channels and rocket nozzle that we pump the supercooled hydrogen through to dissipate the heat. We should probably bring up more hydrogen to dedicate just to the engine. An entire extra nuclear reactor ought to be brought up as well to provide the electricity for the magnet array. And of course, the storage tank for the antiprotons should be back there as well. I think if we get started on the shield wall and trusses, we can put the antiproton tank back there fairly quickly. By the time we collect enough anti-matter, we should be nearly finished with building the rest of the engine. Even if we can't collect enough, we'll still need the trusses for when we figure out something else to use instead."

"So, you still recommend we use the ion drive as well?" Adam asked.

"Oh, absolutely! All thrust is good thrust, remember?"

"Yeah, you may have mentioned that!" Mattie giggled.

"And I still recommend the solar sails. They should help us to at least leave the solar system before the solar wind gets too weak if we decide to go interstellar. We can use them to help slow down at the next star as well. These would be separate from the solar panel arrays that give us our electricity. They're doing that for us right now. When we are between stars, the big solar panels won't generate enough power, so we'll take them down and replace them with the solar sails instead. The sails will provide a bit of propulsion in the way that wind sails do for sailing ships on earth. We'll switch to reactor-fueled electricity inside the station at that point. I guess that's already sort of in the works?"

"Very preliminary, though," Awamila answered.

"Right!" Azaan answered her. "You know, Awamila also thinks we could build some big hydrogen collectors to put up that could be sort of woven into the sails. I'll bet that Mangeet can figure that part out when we build them. That would augment our on-board supplies of hydrogen ions so we wouldn't need to store as much. That'd probably be pretty cheap to build, relatively anyway."

"Sounds rather brilliant!" Adam admitted. "So, how long before we can start collecting antiprotons?"

"Azaan and I have taken the liberty of starting to map magnetic field lines around the Earth from our lab up here. Computer estimates have given us some pretty sound-looking predictions of where we might find pockets of antiprotons. None of this is anywhere near proven, or even tested. But if they are right, it could only be a matter of months before we have our 50 minimum grams collected. If we get more, so much the better. We've even built the collectors; I think we have 60 of them now. If you give the OK, we can send them all up with our nuclear waste disposal missions. Those Shrikes will already have shielding on board. Oh, wait, I guess we'll need to have the storage tank installed up here, huh?"

"Yeah, I know it's hard to be patient," Adam reassured her. "But we'll start the building processes right now. But let's not get our hopes up too high, and for God's sake, be careful every step of the way. This might be the most dangerous and by far the stupidest idea that anyone's ever had. But I trust that the two of you can pull all of this off safely, so let's do it. The other big concern is secrecy. I have no doubt that if anyone in the FAA, the Space Force, or any other living soul gets a whiff of any of this, they'll either shut us down or shoot us down. We'll have to keep the others in the dark until right before we light the candle, assuming we ever can. Anyone can decide right then if they want to get off or not. By the time they touch down at Uplift when they go, we'll have already left for parts unknown. In the meantime, see if you can get a feel for whether or not they might want to come along. I have a feeling that they just might. I hope they will. Then again, I hope you all might too. I shouldn't take that for granted. If all of you do, I think you'll have a happy home here where you'll be highly valued. You'll also be the first humans to reach wherever it is we reach. If this all works like we are hoping, you might get to orbit another star within the next 20 or thirty years. If we don't like what we see when we get there, we can turn around and come back again. Maybe your kids, if you want any, will get to spend the end of their lives right back here where we started. Then again, maybe they'll want to stay with us at the far end of the flight. One way or the other, we'll be the first to try it. Anyway, we'll have at least a year to think it all over. By then this station may just wind up being the world's greatest orbital red herring.

"Right," Mattie said, "Except we never brought up any herring, and I hate herring anyway. By the way, Adam, don't you dare ask me to change my name to Eve!"

After contemplating this request, Adam simply replied, "Duly noted! Now remember! Everything we've just discussed is completely confidential. If any of this gets past that hatch, all our work so far will have been wasted. Nathaniel would probably even confiscate this station and shut us all down. Oh yeah, one more thing, I think this ship needs a better name, especially if we really take it out of port. Why don't you all think about it? We'll christen her at such time as we depart. Oh, and by the way, our FAA inspectors will be making a visit two days from now. I think we'll pass muster, but no loose lips, please." Adam then stifled his spontaneous flight of ideas and adjourned the meeting.

These first weeks gave the crew the first taste of what a lengthy stay aboard this unprecedentedly large space platform might be like. Just getting around the station was daunting enough, especially for beginners. The newer arrivals were at least spared the ordeal of finding that the layout would change almost daily during construction. During this building period, rotating portions of the ship might suddenly stop rotating, causing sudden weightlessness when there had been at least nominal G-forces a second before. When the bridge was finally in operation, at least these changes could be anticipated and the announced a little in advance. But it was all still rather stomach churning. Additionally, there were frequent changes in G-forces when passing from one part of the ship to another that would always persist, even now that the construction was completed. That corridor from the big wheel to the small one was the worst. At first, the front wheel rotation was simply halted when people were passing through it because it was so hard to negotiate. You had to grab a set of handrails to lower yourself "down" the ladder to get "up" to the more peripheral areas. Eventually, everyone learned the precise maneuvers required, but it remained a complex process. Also, the pods of the small wheel were the last to get retrofitted to perform the internal rotation that would be needed if constant acceleration were ever achieved down the line. The center disc around the hanger area was full of the usual square rooms and storage areas you would see on earthbound ships at sea. Rooms like the bridge were arranged to accommodate rotational inertia, although it was unfortunately rather weak inertia. If this ship ever had to haul ass suddenly in a forward direction, you and everything else in the room would be squished into the rear bulkhead, and your day would be completely ruined. These problems would need to be addressed sometime in the hypothetical future.

So, the first two weeks or so after arriving aboard for anything more than a casual visit would generally be spent just trying not to lose your most recent meal. Dr. Annison was always quite busy dispensing Meclizine, both to herself and to the other new arrivals. Those who already had experience with spacecraft weren't as badly affected, but this ship had gravitational challenges no one had ever really seen before. It seemed that with each new arrival, there was a demand to either increase or decrease the rotational rate of the big wheel until a general consensus was found that everyone was fairly comfortable with. Even when a close approximation of one G's simulated downward force was attained, there was still a faint feeling of lateral movement toward the direction of rotation. With such a large ring circumference, the rotation rate was fairly slow, but it was still perceptible. Like going from land to a seagoing vessel, one had to earn one's sea legs, or "space legs" in this particular case. Thankfully though, everyone eventually got to where it faded out of their conscious perception. Paradoxically, once you finally really felt comfortable, you actually started to dread the idea of going back "down there" to Earth. Adam and Mattie, plus the others with extensive spaceflight experience, were the first to really embrace the idea of staying aboard exclusively whenever possible. This had apparently been Adam's ideal from the very beginning when he first envisioned the station. He was now delighted to find that it was entirely possible for him to live up to his vision. He was quite exhilarated by this realization. Mattie had already found herself caught up in his enthusiasm by the time they fell in love. She therefore never really bothered to get too fond of the wealth and influence she could have had if she had stayed on the ground. Extreme affluence was fun and all, but she never thought of it as a permanent condition. As each task force member was drawn into the vision, they too became gradually more committed to the idea of living their entire lives on this station away from their home planet. To be fair, it was doubtful that they could have really thought that things would actually have come this far so quickly. Even the new medical people, though not yet feeling particularly welcome, were never recorded as voicing regret for having chosen this bizarre new lifestyle. They were all aboard now, and they appeared to all be on board as well.

The more veteran members of the crew (by only a few months) all made it their number one priority to be helpful, enthusiastic, and accepting of all of their new crewmates. Mattie, being "First Mate," was usually the first person awake every morning. At 6 a.m., she made her way to the bridge to relieve whoever had been on watch that night. She would generally find that everything was in order and then transfer a copy of the day's duty roster that she had made the evening before to her cell phone. She then broadcast it to the crew and asked everyone to acknowledge that they had received it as soon as they could. She checked the status of all ship's systems and assigned the appropriate people to deal with any development that required attention. According to the overall plan she was developing, everyone started their day at 7 a.m. and presumably got themselves cleaned up, squared away their apartments, and prepared their breakfasts. On some days, a communal breakfast was prepared in the kitchen area in service corridor one, with Sharon the main cook. In actual practice this would be done only once a month or so, per crew preference. Ideally, every crew member was to acknowledge receipt of their assignments by 8:30 a.m. and begin their workdays by 9 a.m. The medical staff went to their particular pods to do status checks and inventories and report any problems or needs back to Mattie. Initially, completing those initial physicals and medical histories were their main assignments. Grumbling aside, no one was delinquent in keeping their appointments, which was a good beginning. Everyone had deduced that Mattie was a bad person to cross, and she always seemed to know if you weren't adhering to your assignments or appointments. On this crew, however, you wouldn't be aboard if you hadn't shown the capacity for absolute dedication. Therefore, Mattie was a perfect person to supervise them. Overall, though, everyone appreciated Mattie as much as she appreciated them. This was seen as a good omen by the people who might eventually elect to lock themselves aboard this isolated ship for an extremely long time.

After the morning departmental checks, the crew would break up into their assigned work groups for the day's assorted chores. Sharon and her group, for the day, would start out feeding the fish and doing maintenance on the tanks. Water temperature, pH, and general cleanliness were assessed. It was decided not to start angling or otherwise collecting any critters for eating purposes until each species had demonstrated that they could perpetuate themselves. The first new little arrivals were among the shrimp, but none appeared on the menu until this first brood of babies grew to adulthood. Next came the inspection of the plants, all of them, regardless of where they were. The exceptions were the plants people had in their apartments, although no one had really gotten too far yet in placing any houseplants in their rooms. Sharon was the exception to this. Watering in the arboretum was quite time-consuming, even though at this early stage, about three months after everyone was aboard, there were only saplings and sprouts of corn, wheat and other crops. For this task, the watering hoses and spouts had to be placed right on the soil to allow the water to disperse via capillary action. Once you eventually floated yourself and the hose to the optimal position, you needed to avoid having globs of water and mud dispersed all over the place in the largest compartment on the station. Finally, Sharon would end the day demonstrating the production methods used to manufacture synthetic meats, usually ending up with enough to potentially feed the entire crew dinner by the end of the day. This early on, people generally chose to let her freeze the stuff for now, then go home to their own prepackaged food. It helped Sharon's confidence that no one ever complained about the grub.

The engineering team spent their days meticulously scouring the corridors and equipment rooms from stem to stern, checking that everything was clean and operating properly. This early on, there was no question of there being any fluid leaks or structural damage. But if there had been, it would immediately be reported, and plans made to either fix or replace the damaged component. The 3D printer equipment was also checked daily, and soon enough, requests were made for Mary to whip up some little article or other for the ship or for individual crew members. For example, in the first few weeks, it was observed that no provisions for haircutting or styling were ever made before the crew settled in, so things like scissors, a hair vacuum system, and a barber chair were built and assembled, and then placed in one of the service corridors. Customers would need to wash their hair in their apartments before reporting to the barber. Unfortunately, no one knew who that barber would be. Volunteers were requested, and they could tell Mattie if they wanted to give the assignment a go. Maybe a robot could be built? Meanwhile, everyone got shaggier by the day. With increasingly flowing manes, the engineering teams spent several hours every day wandering or floating through every corner of the ship, which seemed larger to them than they had originally expected. True to the principle of generalization in crew training, work assignments were shuffled frequently. Except for the team leaders, the compositions of the teams were changed daily to eventually ensure everyone could do routine ship maintenance with the same level of competence. Slowly, everyone gradually became familiar with how this station was supposed to be operated. Except for the medical jobs that only the medics could do, the crew was cross-trained so that even fairly complex technical jobs could be performed by every person aboard. The other exception was the Physics lab, which was the sole responsibility of Awamila and Azaan. If anyone wished to apprentice with the docs or the physicists at some point in the future, it was their own responsibility to arrange it.

In regard to everyone's daily routine, time was allotted for everyone to manage their own lunches between 12 and 12:30, and all routinely assigned work was supposed to be done by 5 p.m. If the teams were done with their assignments earlier than that, they were free to go and could use the remainder of the day as personal time. This usually meant returning to their own apartments to do whatever they wanted. Adam was anxious to convey the idea that crewing this ship was essentially just a job as anyone might have on Earth. You simply went home at the end of the workday like everyone else. The goal was to convey that this way, you might not feel homesick as much. It was altogether too soon to find out if this was an effective strategy or not. Whatever the case, people seemed to remain relatively content after these first few months. There developed a sort of everyday community social behavior in the evenings as well.

Friendships were starting to form, and no one was proving to be reclusive. The leisure areas were being used on a daily basis. As Adam had been dearly hoping, a few romantic pairings even seemed to be forming. He knew that if the crew was to function as harmoniously as a crew would need to, it would probably be a good idea to come up with some sort of formalized code of behavior and expectations. He felt that all human communities needed these kinds of standards to function smoothly over time. He envisioned that the crew should enact an actual set of laws and guidelines for this new little community that would, in truth, be a separate planet all its own. Standardizing such rules would be an almost mandatory requirement if this audacious idea of his was to reach fruition. But for now, this idea needed to be put on a back burner. There was so much more to do at the moment. At this point, it wasn't even clear whether this would be the final group of inhabitants to occupy this particular little ship.

It was also discovered that as long as the station was orbiting the Earth, there would be a constant demand for the crew to entertain visitors. These were mainly wide-eyed influential tourists, who arrived in small groups two or three times per week. They came up by the Shrike load, but thankfully only in groups of two or three visitors at a time. They docked at the main port of the small wheel and were ordered to stay in the old hotel room pods with the puny gravity. Just after principal construction was completed, some of the first visitors included Nathaniel Floatingfeather, along with the two current FAA inspectors in tow. By this point, Shrikes visiting the station had now been converted to use the same mixture of Nitrogen and Oxygen used aboard the station itself. It was also maintained at the same pressure as the ship's. This made these specially outfitted Shrikes seem basically the same as earth-bound passenger liners. This made them more accessible to civilians. The days of denitrogenating in a pressure suit were now over. Only those preparing for an EVA outside the confines of a space vessel would need to rebreathe pure Oxygen for an hour or so before donning their bulky EVA spacesuits. Visitors could now move freely between a Shrike and the station interior.

Since Uplift was an American company, and the research station spent a part of every orbit passing over American Airspace, the FAA invoked government privilege to send inspectors aboard. It was initially agreed that only one of these visits per year would be made, but somehow Nathaniel increased this to once every three months. Adam was aware of this, and played along, knowing that between these inspections he could do pretty much whatever the hell he pleased on his own station. The "facility" had never officially been declared to be US territory, and the most recent codes of international law weren't really clear about national domain questions for spacecraft. Nevertheless, Adam thought he ought to keep up appearances and present the station as merely a US sponsored research facility with the harmless sounding aim of seeing how long a space station could operate in orbit without assistance from the ground. At this point, in reality, this truly was its only purpose. Nathaniel, however, knew Adam well enough to have suspicions that much more was afoot. He met alone with him on the bridge at lunchtime while the inspectors were off at the kitchens with Sharon conducting an inspection. He asked Adam for a frank summary of his immediate intentions.

"Well, since you asked," Adam said. "I think you realize that I intend to stay aboard up here for as long as the mission lasts, regardless of how long that is."

"Yeah, I kind of gathered," Nathaniel answered. "It's not like I've not been paying any attention to your business plans since the first day you took over. You've been running Uplift from up here for the past couple of months. Is this going to be permanent as well?"

"Sure, why not? Lots of CEOs tend to work remotely. It seems to be working out OK, doesn't it? You know you can call me anytime, and you have been."

"True enough. I also know I can call you down if I really need to. So far, I haven't. What I really need to know is whether or not you ever plan to, shall we say, get out of the neighborhood. I'd be nuts not to think you may be thinking along those lines. If you research how to build a ship that can skedaddle and stay gone for a really long time, then actually build it, then why not use it in the way it's supposed to be used?"

"Mainly because we don't have any engines," Adam replied truthfully.

"Then what's Azaan been doing all this time down in that little basement of his?"

"I imagined you would have known about that by now. He's doing just what I asked him to do, that being propulsion research. Why else would I have hired him? I also imagine you know we want him to move some those projects up here to work on. What's the point of research if you can't test your theories in real world conditions? Sorry, off-world conditions?" The men both laughed a little. "I hate to put too fine a point on it, but where else could we go to do any meaningful testing? All sorts of research entities have talked about doing what Azaan has been doing, but no one has ever gotten as far as he has. Except maybe with conventional rockets! This isn't really US territory anyway; what authority does anyone have to stop us?"

"That depends on whether or not you're messing around with nuclear material. If you are, we have a whole world of authority. If you want to see which world, look out that window over there."

"Nathaniel," Adam confided. "You're virtually my father now. I'll level with you. Azaan and I were tinkering with the idea of nuclear propulsion. We came to the same conclusion as you a few months ago. Flat-out nuclear propulsion might be developed someday, but not by us. Right now, we'd like to bring up an Ion Drive engine just to see how it behaves. We'll be using Argon as our fuel source. We're also planning to test some solar sails. Big ones, big enough to push around this huge beast you're sitting in. God knows where it'll end up leading us. If we decide to try leaving orbit, we will. In that case, you get the company for yourself. And everything I leave behind too, by the way. At least until I come back, and Mattie and I are both alive when we do. In that case, you'll have to give it all back."

"What are the odds?" Nathaniel asked.

"Pretty highly stacked against it, now that you mention it. But I never thought we'd be sitting in a spacecraft like this so soon, either. Who the hell knows?'

"OK, I think I at least understand what you're thinking in that immensely unorthodox brain of yours. Can you promise me you'll actually try to use your best judgment from here on? Please say you'll at least do that."

"No problem," Adam said in a sincere tone. "My problem has always been that my idea of good judgment tends to diverge a bit from everyone else's. I will certainly promise I won't do anything that is known to be illegal when I do it."

"I guess that will have to do then." Comically, it suddenly registered that Adam had just offered to make him the wealthiest man on Earth. He choked on his next words. "Just to clarify, did you say *all* your stuff?"

"Yeah! What use would I have for any of it if I wasn't even on the planet?"

"Hmm, that is a very good point," Nathaniel said, still feeling rather flustered. He recovered quickly. "Well, I can't say I haven't suspected something like this. Just a word of advice, though. You know that pharmacist of yours is still active-duty Space Force, right?"

"Sure! I'm the acting Human Resource guy up here, you know. I realize he could rat me out pretty quickly if I do anything too naughty. Then again, maybe he'll just consider going along with whatever we do as part of his Uplift job description. Whatever he decides, he's quite welcome to scoot downstairs to Baltimore whenever he wants. Unless he waits too long to decide, you know? Then he's coming whether he likes it or not."

"Don't tell him until just before you leave, that's my opinion. But be sure he's serious if he wants to stay. He'd be a pain in the ass if he got stuck here against his will."

"Yeah, I think that's kind of the plan," Adam confided. "But hell, we don't even know what will actually wind up going down. We'll just have to bide our time while Azaan fiddles around. Just keep those FAA suits away from us, please. That basement facility of Azaan's will be shutting down soon. Then everything will proceed from up here. Ok with you?"

"Yeah," Nathaniel said." Geez, why did I ever decide to let your dad stop being a janitor?" Nathaniel asked.

"I don't know," Adam said. "But if you need a job, we sure could use a clean-up guy up here."

"Seems tidy enough to me, or is that just for inspection? Well, that's your problem, I suppose. Now, what's for lunch?"

After this conversation, Adam satisfied himself that Azaan should indeed begin bringing his machinery aloft as soon as possible and begin construction of the aft propulsion systems. Azaan was already prepared for this green light. He would stick with the excuse of "research" if anyone started asking what he was up to. The Shrikes loaded with the parts for the radiation screen started arriving the very next day. With Azaan's gift for efficient planning, the process would only take about two weeks to complete. The large screen would have the added beneficial side effect of obscuring all the subsequent work from being viewed by anyone in the Arboretum or the physics lab. At the next full crew meeting, Adam explained to anyone not fully on board with the station objectives that routine propulsion research would soon be starting. But it was, he equivocated, only research. This was all done mainly to satisfy any curiosity that Kevin St. Patrick might entertain on the Space Force's behalf. Communications from ship to Earth were not yet subject to any sort of censorship from Adam. That precaution would eventually be imposed but hadn't been yet. Kevin could easily alert his superiors if he felt the need to play snitch if something seemed a bit amiss. Adam chose to ignore the risk and simply play dumb. For the time being, the crew continued to spend their days perfecting the routines of shipboard life and, more importantly, discovering how to simply just get along with each other. This was, after all, a very diverse crew coming from a wide variety of backgrounds. The original Uplift people already knew each other well, but they continuously needed to be reminded that a few others were not from Uplift. They, therefore, needed to take time and patiently explain themselves and their ways to one another. Adam's guiding principle of conduct was tolerance. He repeated this again and again. The one mistake you could make that would attract Adam's evil eye was to express any word or commit any deed that might jeopardize harmony on board the station. Arguments were to be suppressed in exchange for compromised solutions whenever possible. No derogatory or degrading comments were to be expressed, even in casual speech. These sorts of expectations were, of course, rather contrary to usual human norms, but the captain adamantly believed them to be absolutely essential if the ship was to run smoothly. The crew was told to bear this tolerance in mind at all times. Every crew member was also expected to call out anyone who violated these basic rules. He needn't have worried though. Every single crew member had come from a profession where good behavior had been an expectation throughout their previous working lives.

Weeks passed by, and bit by bit, each individual now aboard steadily became familiar with their new chosen lifestyles. Some accommodations began to be made by the section chiefs to lessen any work expectations that were found to be unnecessarily stringent. Uplift jumpsuits had been the unofficially prescribed mode of dress for everyone, and most of the crew found them comfortable. But eventually, a few crew members were heard complaining in the background that uniform dress requirements seemed like an attempt to squelch out any individual identity you had left when you joined the crew. It made it seem to them almost more like they were joining the military, and this was not intended to be a military-sponsored project. Adam listened to these arguments and realized they were completely valid. He reversed the policy and decided that as long as no frankly unsafe clothing was being worn, people could send down their own clothes and have them shipped up to the station. Soon, fashion choices from all the nations of crew members' origins were in evidence around the ship. Usually, these were only worn after hours, as the work jumpers were far more utilitarian during the messy laborious days. Manjeet had been the one to first stretch the rules when he had insisted on wearing his blue turban as soon as he first came aboard. Since that time, he had never once been seen anywhere on the station without it. No one could stop him if they tried, of course, but no adverse effects had ever occurred as a result of his wearing it. This called into focus the need for tolerance of everyone's cultural rights here in this enclosed little space. When first coming aboard, everyone on the crew had been told they could take at most twenty articles of personal clothing, twenty sets of undergarments, and four pairs of shoes. In an effort to appear enthusiastic, everyone initially complied. A large crate of jumpsuits in various sizes had been shipped up to the station very early on in the construction process. These suits were unisex and actually disposable, but they were machine washable as well. When washers and dryers were installed, it was obviously more desirable to keep reusing them instead of continually tossing them out. Bio-tolerant detergent was used in the laundry so that wastewater could be filtered and the impurities used to fertilize the plants. Once the policy was changed, though, it became more common to see people occasionally choosing to wear their Earth clothes if their duties were not expected to be overly messy. Awamila generally preferred to wear the silk-based clothing that was the standard of dress in Pakistan. This immediately made her the best-dressed member of the crew, and no one ever found cause to raise any objections. Tariq al Abdullah, the neurosurgeon, also gradually showed a preference for the clothing of his homeland, but the wearing of traditional robes proved to be somewhat hazardous in zero-G areas of the ship. He eventually moved to use the jumpsuits along with traditional Arab headdresses, which were no more inconvenient than the long-flowing hairstyles preferred by Mattie, Sharon, and Victoria. The other females either cut their hair or wore it in a bun.

These slight cultural differences also revealed the need for tolerance in the matter of religion. Awamila and Tariq began performing their prescribed daily prayers immediately upon arriving on the station. About this time, Awamila also assented to be called, more simply, Mila. (This seemed to be a pet name Tariq had come up with for her but was not a nickname in general use in Pakistan). They were both practicing Muslims, although religious customs in their respective countries were anything but identical. Initially, facing Mecca when praying was a bit of a challenge for the two of them, and they designed an app for their cell phones to compute which direction they should face based on where the station was in orbit at that particular moment in relation to Saudi Arabia. Eventually, they decided that if their heads were positioned facing as close as possible towards the Earth, this would be sufficient. In accordance with Adam's general policy, no one ever once objected to their interrupting their duties for prayer time. In fact, Juni programmed the onboard computer to broadcast the call to prayer over Tariq and Mila's cell phones at the appropriate times of the day. These could be heard by anyone nearby either of them unless the two happened to be in their quarters that day. This sort of paging service had long been available on board all contemporary vessels at sea owned by Muslim nations. It soon became a reliable way for the station crew to gauge what time it was. Praying also appeared to draw Mila and Tariq closer together, which was always rather sweet to see.

As for religion in general, several other crewmembers practiced the Christian faith, and they occasionally met in one or the other's apartments for makeshift worship services. Manjeet and Juni obviously practiced Sikh religious customs one day each week and were exempted from crew duties when doing so. Equally important, those members of the crew who did not identify themselves with any religion at all were to be treated with the same level of respect and deference as those who were. An initially unwritten agreement was made and enforced throughout the crew that specified that no one belief system was ever to be regarded as mandatory aboard the ship, and no one's personal faith was to be treated as superior to the others.

As long as the common civil laws of all nations were enforced, then every crew member would remain in good standing with the others. For example, no theft, assault, blackmail, intentional murder, or other universally unacceptable crimes would be tolerated. A true legal code would be devised later, and some means of punishment would be agreed upon if it was violated. At a minimum, no religion-based discrimination or bias would be allowed to exist on this ship. By this time, this was also the earthly standard for all naval and sea-going vessels on the world's oceans. Such codes were known to be absolutely essential for any earth-bound crews to function efficiently. Adam, as it happened, had come to this conclusion on his own pretty early in life. His father had never committed to any religion, and Adam never attended any church on his own accord, with the exception of the occasional wedding or funeral. Even his own father had a simple, non-denominational funeral. He respected the validity of all major religions practiced down on his home planet.

Another social custom that may or may not have been beneficial for ship-board life was the consumption of controlled substances. Adam, always the social outsider, had never liked alcohol or any other sort of recreational drugs. This wasn't really due to any moral objections overall. He just didn't like the way his body felt when under the influence. He really hated the idea of spending inordinate amounts of money for the purpose of feeling so crappy. Therefore, he didn't personally imbibe. He didn't much care either if anyone else did. It simply was none of his business. As long as it didn't harm anyone in the process, he had little to say about it. That philosophy was largely applied by him in regard to Mattie, who didn't mind a drink or two in the evening when she could get one. He and she both shared the same feelings about marijuana. To them, the infamous weed was pretty much just the same as alcohol. But as far as anyone knew, all there was on board were alcoholic beverages. On the other hand, there were probably numerous occasions when cannabis might have made it onto the station without being detected. Time revealed that there were varying ideas on the subject of recreational substances by everyone aboard. Everyone agreed that the smoking of marijuana was an exceptionally bad idea, though. Even with the atmosphere aboard approximating that of Earth, burning anything aboard an enclosed vessel was too dumb an activity for a highly intelligent crew to attempt. Sharon eventually made it a priority to include hemp in her list of cultivated plants, but these would be used to make orally administered recreational variants only. Even these would be highly monitored, and the crew was advised to report anyone showing signs of intoxication while on duty to higher authority. Prudish Adam had to be badgered quite heavily to agree to bring any sort of supply of alcohol aboard whatsoever, but he was gradually convinced to keep a modest supply of beer and wine aboard ship in perpetuity. Sharon agreed as well to grown grapes and other crops which could conceivably be used for eventual production of alcoholic beverages. Methods of producing this hooch were also loaded into the computer database. Adam agreed that in time, with good behavior in this area, production gear for homemade booze could conceivably be printed up by Mary. At this early stage, however, everyone was simply too busy to think much about imbibing in anything at all.

Nobody was ignoring their recreation though. MD Peter took it upon himself to pester all his crewmates to make certain they attached themselves to a hobby of some sort. He recommended that whatever distraction they chose, it should be practiced on a nearly daily basis. He himself led evening jogs around the big wheel every weekday at 6 p.m. Mangeet, not fond of running, concentrated more on weight and resistance training in the gymnasium. Sharon took the liberty of ordering up a sewing machine, loads of fabric and thread, buttons, zippers, scissors, and anything else her quilting friends in New

Mexico could scavenge. The first ever quilting club in space, she hoped, might one day become a reality. She also offered to teach classes in knitting and crocheting. Adam was the first to express interest. No one had so far begun to try and learn to play any musical instruments. But Adam was a hardcore rocker, at least in his own mind, and music blaring at fairly high volumes in the evenings before bed was identified as coming from Thorne's pod. Rather than eliciting complaints, instead the practice gradually spread throughout the habitation wheel. This was allowed only in the evenings, and rock and roll did not monopolize the music style. No big parties occurred until the autumn holidays came around. Starting in November, festivities devoted to a variety of the world's special occasions were worked into everyone's schedules by Mattie. This was the time when the individuals on the crew of the Uplift Orbital research laboratory began to really function as a unit. Everyone seemed to feel the change, and Adam made note of it in his log. Adam was fond of unofficially calling the station a "Village" in order to instill a feeling of being an Earth community aboard their space bound habitat. People worked their jobs in this village and lived there too. This village was just a bit out of the ordinary, that's all. It was small and rather cramped, but there were worse places a person could live. They found that they could make the place a happy home, and that mattered a lot.

A few of the villagers were doing a bit of clandestine moonlighting in addition to their usual day jobs. Gregor and Manjeet, along with the four robots, had finished the big radiation shield at the back of the station. Everyone had been told to expect this as part of some experimental research soon to be conducted beyond the shield. This was done as part of the subterfuge done to keep the non-engineers from passing gossip down to their friends below, as has been discussed before. In the month that followed, unbeknownst to the majority of the crew, various fuel tanks, pumps, and an array of heat radiators were installed beyond the shield. Power from the solar arrays was routed to the area to allow for the installation of a much larger 3D printer in a compartment attached to the frame trellis growing beyond the shield. This was to free up the smaller printer in one of the big wheel service corridors. It was also necessary because getting large, fabricated structures from the front of the barrier to the other side would have been horrendously complicated. There was soon enough trellis area to accommodate the containment vault for the new and mysterious cargo of antiprotons that Manjeet and Gregor planned to begin gathering almost immediately. The curious crew couldn't help but notice the huge uptick in Shrike arrivals back behind the radiation shield. Those in the know were instructed not to reveal any knowledge they might have about what, specifically, was going on back there. It was just to be described as the installation of equipment to research a wide variety of propulsion possibilities. About two months in, Manjeet announced that he was needing to bring a new crewmember aboard to help with this research. Jeter Hamm, a Norwegian who spoke English, held a PhD in Nuclear Engineering. Azaan had instructed him to tell everyone only that he was added to the crew to help assemble the ion drive that was being built to help him test the "feasibility "of its use on a vehicle operating in a vacuum. No one seemed especially convinced by this excuse, but no one raised any particular objections either. Awamila (Mila), who would know better than anyone that this assertion made no sense, was of course one of the few people aware of the anti-matter research about to be conducted. One project that was begun in front of everybody were the huge solar sails now being attached. These were actually to be added to the periphery of the solar power panels already in place and unfurled when a certain distance from the Earth had been attained. Gregor had advanced his ideas even further. Not only would the solar winds push the station forward when striking the sails, but the sails would begin collecting sparse interstellar Hydrogen in both atomic and ionic forms. Both of these particles would be very useful in ongoing ship operations that would hopefully last for quite a long time. Using Shrikes and bots, the sail installation was one of the simpler construction tasks to be undertaken on the station. It was also the quickest.

One highly surprising development that came within a month of the installation of the antiproton tank was that Manjeet's ideas about the collection of these elusive particles were proving to be correct. Down on the ground, missions devoted to disposal of spent nuclear fuel disposal were beginning to be outfitted with the special collection devices that had been built in the previous few months. These small canister-shaped collectors were loaded into fairly inconspicuous launching devices. These were deliberately built to exactly resemble the containers holding the expended uranium and plutonium soon to be shot off into deep space. 40 to 50 of the collectors were to be gradually fired off to penetrate regions assumed to be fertile with antiprotons due to their expected presence in the magnetic strata of the Van Allen belts. These containers had their own little maneuvering rockets, which could propel them through the belts, collect the antiprotons, and then guide them back to the vicinity of the Uplift station. Here, they could be retrieved by other Shrikes doing apparently innocuous supply runs. They could also be grabbed by the robotic arm now in place at the rear of the station. However they arrived, Sparky had been programmed to catch them with his hands and then plug them into the supply tank to discharge their very unusual contents. They would then be available for reuse in future Shrike missions. The particles could be stored in this tank for an indefinite period. At least that's what all the projections said. The special Shrikes all were instructed to maneuver to the rear of the station before deorbiting so that Sparky could load any empty collectors back into the cargo bays. The FAA inspectors on the ground, being used to inspecting these radioactive material containers only for escaping radiation, never seemed to find anything unusual going on. Manjeet was astounded to discover that for every 50 collectors sent out, as much as 75 mg of antiprotons could be collected. With a total necessary goal of 10 grams, or 10,000 mg, that would mean around 134 such missions to reach their goal. Adam contacted Nathaniel to request an uptick in the number of supply runs to the station. These were now to be combined with the waste disposal flights "to increase efficiency." Nathaniel protested, since this would conceivably risk contaminating the supplies before they were brought aboard. But Adam assured him that everything passing through the air locks would be tested with Geiger counters before being brought in. The nuclear waste was in fact put in highly secure containers before being inserted inside the little rockets that would shoot them off to the planet Mercury, and no radioactive spillage was ever detected. Even so, Nathaniel would always be suspicious of Adam's request. There were already about 40 runs each month, so the period involved would hopefully be around three months, give or take. Strange requests were well within what was normal for what Adam might ask for from orbit up to this point, so Nathaniel went along with it. Heck, Adam even asked for a shipment of earthworms once or twice. This was in March of 2088.

Coinciding with the clandestine components of the Shrike flights, construction of the actual propulsion systems continued apace. Truss systems began growing to support an intricate and extensive group of conducting tubes, substantial magnetic field generators, and heat management systems. Actually, two ion drive units would be placed on trusses peripheral to the future central antimatter engines. Since all thrust is cumulative, the two types of engines would usually be operated simultaneously unless certain maneuvers required something else. Each of these propulsion systems required massive amounts of electric power. It was felt that the only possible solution for generating that electricity was nuclear power. That in itself was quite problematic, as putting nuclear material in Earth orbit was about the most illegal activity in which one could engage. Even if this antimatter matter could be brought to fruition, the nuclear matter matter might be even harder. Adam, unknown to everyone save Mila, was working on a possible solution to this challenge, but attempting it seemed at the time to be nothing but the worst sort of folly. But the days ticked by, and shipboard life ticked on as well. The main challenge remained food production, and this could only be accomplished with pure crew manpower and attention. Sharon Layton was happily proving herself more than up to the challenge. She lived up to the assignment to be the de facto first resource for all of this activity. While all of this unprecedented Engineering achievement was happening behind the scenes, Sharon became a sort of mascot for everything else that was happening aboard the station. Adam couldn't have been more grateful. More than anyone else, he was the one who constantly kept in mind the fact that this outlandish mission he had dreamed up was actually the ultimate test anyone could ever dream of imposing on a group of human beings. The humans were the true critical component in this endeavor. The means of actually transporting them was only secondary. Maintaining the crew's happiness, health, and basic humanity became Adam's true focus, even though no one seemed to realize he was doing it. The crew could deal with the hardware part of things. This philosophy of Adam's would become evident in his later communications with his home planet. But for now, he tried to come off as a sort of aloof management type who seemed to be more interested in running his business from home than any trying to appear like some significant visionary. He generally was not out there helping building stuff or watering the plants. That would come later. Just now, he spent most of his days playing captain on the bridge, and his nights snuggled away with his darling Mattie as any good working husband would. Not that he was disliked by anyone in any way, far from it. The crew probably often felt like they were missing out on his energy and his attention. This was not to last much longer though. As Adam came to feel like all his vision and effort might just start to pay off, his demeaner became more interactive and enthusiastic. He now was beginning to feel more like the head of the household than a mere administrator.

Chapter 14
Road Trip Anyone?

From the ship's logs and meeting minutes transmitted in May 2088...

The holidays of 2087 were a period where the crew of the Uplift Orbital Research Station had truly begun to feel settled in, at least in terms of the day-to-day operation of the ship. Various crops were ripening and being used for the first time. Pumpkin pies could be baked using the ship's pumpkins, not canned ones. Wheat and rye were available for making bread. Sharon had taught a class in breadmaking, and there were plenty of eager students. That Thanksgiving, the first big frozen turkeys were defrosted and roasted in Sharon's large ovens. For those who did not wish to partake in Western holiday traditions, other seasonal feast foods were available and served as requested by the feasters. In actuality, no one objected to celebrating traditional American holiday customs, as they were ostensibly on an American vessel. Perhaps holidays from other cultures would be observed in theoretical future years. By the December holiday season, one of the small spruce trees in the arboretum was decorated, and the whole crew was invited to an improvised zero-G picnic. The tree wasn't actually cut down. Adam called everyone together and had Mattie schedule a whole crew meeting in the conference room area of service corridor number two. This would be held on the second of January at two p.m. He announced that the subject of this meeting would be to "realistically" plan for how the station might actually leave Earth orbit for a trip into deep space. They should try to envision a trip that would last "for several years" or longer. This was all hypothetical, of course, or at least that was the implication. But since the crew had by now demonstrated that their station seemed quite capable of sustaining itself for an extended period with the supplies and equipment currently on board, it seemed time to expand their thinking towards what might come next. He asked everyone to seriously consider how they might feel about actually taking such a trip. They were also asked to brainstorm about what still might be missing from their plans that hadn't already been considered. This was not just about technical issues but social ones that might need to be implemented during their envisioned onboard lifespans. Things like marriage, starting families, aging, and interpersonal relations aboard this minuscule self-sustained ecosystem should be the focus of their thinking, and they should share their conclusions at ongoing crew meetings. This was part of the original task force mission, and the time had come to start asking these questions. The transcripts from these meetings will no doubt be studied by aerospace planners for generations to come and used as a basic blueprint in future mission planning. Here's what was discussed.

January 2, 2087 – Adam called this meeting to order using his accustomed format. The meeting area was a dedicated space in the second service corridor ring. This had more stable simulated gravity and had more space as well as better audiovisual equipment. There was a table large enough for the entire crew, with the floor facing toward the outer edge surface of the ring. A large flat screen was connected to the ship's main computer would aid in any presentations that might be needed. Everyone had brought their laptops, which were also connected to the mainframe.

"Well then," Adam began as usual. "I know you all have been considering the questions I posed at Christmas. I've heard you all muttering under your breath about it all week. Now remember, let's start on the assumption that this is all pure conjecture. But every long journey generally begins with nothing but a dream. Obviously, we can't just leave the planet of our origin on a whim. But on the other hand, at least we almost have a vehicle we can use to make such a trip. That is, assuming we can have an engine to make a journey with."

"And do we have an engine?" Kevin St. Patrick asked.

"Well, theoretically, we have several," Mangeet responded. "Not everything has been built, though. Besides, there's no sense in even trying if we don't have a destination in mind."

"Fine," Mattie spoke up. "That seems a good first question to ask. Where should this little road trip take us?"

Adam called up a blank screen behind him. "I suppose we should start in our solar system," he said and typed. "Mars has already been reached by others. I personally am not interested in merely following in someone else's footsteps. It would be a waste of our superior technology, for one thing. And actually, all the other planets are pretty much uninhabitable. How about moons?"

"Well, there are several potential candidates around the gas giants," Mila answered. These will only be a couple of years out and the same back if we use things like solar sails or an ion drive. All of these moons are frozen. Some may have liquid water under a thick ice crust, but none has an oxygen atmosphere. If we brought breathable air down with us and made enclosed habitats, we could do some exploring and stay for a little while. I guess we could bring some Shrikes along and fit them with skis or something to land and take off. We'd also need to take down a 3D printer to make shelters and storage areas like they did on Mars."

"I'll admit, no one has explored any of these moons in person yet or has any plans to as far as we know." Adam agreed. "Still, if we have a spacecraft capable of operating for much more than three or four years, it seems to me a waste to use it that way. Just for laughs, what targets are out there a little further?"

Mila now took over the conversation. "Nothing just a little further, but if you don't mind a few light years, there are several in the vicinity."

"Such as?"

"Well, everyone's heard of Alpha Centauri, right? Of course you have! You might even have heard that it's actually a star system containing not one, but three red dwarf stars. Red dwarves are by far the most common sort of star in our particular galaxy. They are much smaller than our wonderful local star here in this solar system. On the other hand, their life span is hundreds of times longer than our sun. Maybe trillions of years. Two of the Centauri stars orbit around each other. All three have planets. One of those mutual orbiting suns is a G-type star like ours. The closest known planet is Proxima Centauri B, but its sun is much less massive, a red dwarf. Red dwarve planets have to orbit much closer and circle their stars at much faster rates because the stars themselves are much smaller. We're talking years lasting for only Earth-length days. It appears none of the planets rotate either, so one side is always pretty hot, and the other is likely frozen solid. It depends as well whether any of these planets might have magnetic fields that would deflect the radiation thrown off from all three stars and thus allow for a thick enough atmosphere to develop. Otherwise, all the air would be blown off into space. See how lucky our little Earth is? In their favor, all the planets we have detected in that system appear to be in the "Goldilocks Zone." That's what we call the area where surface temperatures would be in a range that wouldn't be immediately lethal, not too hot, and not too cold. They all also seem to have a mass that nearly approximates Earth's."

"Don't you think it would be worth a look?" Adam asked.

"As a physicist, I think approaching that system, slowing down, and threading ourselves around those three suns to examine which planet is best would be a challenge in itself, even with our supercomputer helping us out. Still, most experts seem to think it's the best choice. Now, of the three stars, the G-type star might be the best bet, but we know too little about its planets.

"So, it seems you're not one of those experts who support going there. Am I right?"

"Well, there's one interesting system called Tau Ceti. It's 11.89 light years away, just below, or south, I guess, of the plain of our solar system. It has a G-type sun and five known planets, and one is certainly in the Goldilocks zone. As far as its composition, there is reason to believe it has water on the surface."

"Any other stars you like?"

"There is the Trappist system. It surrounds a red dwarf with 5 planets in the habitable zone, but all with years lasting less than one Earth week. They're probably all tidally locked too. That is, they don't rotate. Biggest problem here is that it's just over 40 light years off."

"What do you think would be the fastest we could get our ship moving, in your wildest estimates?" Sharon ventured to ask.

"At best, about half the speed of light, employing a gradual increase in velocity that could simulate one G," Azaan answered. "23 or so years to Tau Ceti, which seems realistically doable if the engines operate as intended. It would only be more like eight or nine years to the Alpha Centauri G star, but we don't have as much data about what planets might be orbiting it due to it being a three-star system.

"And what makes you think you can achieve that kind of velocity?" Kevin asked.

"I have a few different spoons in the pot right now," Azaan replied." I won't know for a few months or so at least if there are any truly viable options." Adam and a few others knew this was not at all true, but they also knew Kevin was not the one to tell about it. "I can say that we are ready to install our ion drive," Azaan continued." This can really get you going pretty quick, but it takes a few years to build up the thrust to a significant level. We'll also be installing solar sails alongside our current solar power panels. This would act in the same way, very gradually. But they would also slow you down as you approach another star."

"Well, if you truly want to go, and it takes that long, I guess we should think about starting when we're still young," Peter Hosokawa said.

"For what it's worth, I say we try Tau Ceti, and get those supercomputers going trying to figure out the fastest way to try it!" Sandra Annison blurted. She was not usually that blunt.

"The computers are already working on this and a few dozen other options too," Juni confirmed. "They have been since before we even moved up here."

"I was hoping you'd say something like that," Kevin said. "That should save us some time."

"Us?" Adam said. "Does that mean you'd actually consider participating in this hare-brained scheme? I mean, really? I'm glad to hear it! Seriously though, I think this is the time for all of us to decide whether or not you'd stay aboard if we found a way to get out of here. You've heard the timeline estimates. It's still a long shot, but I want to know who here is thinking about committing to something like an interstellar journey and who isn't. If it isn't you, I need to have the time to find someone who can take your place. So tell me soon."

"I'm still active-duty Space Force, and I know that you know that. I can say that since I was assigned to work with you on this project, my superiors agreed that they should eventually consent to let me to continue my assignment here for as long as it took. I told them I was fine with that. I don't really have anyone on Earth that I'm leaving behind, so it's all the same to me. As long as I build enough confidence that you won't just blow us all to hell the first chance you get, I'm with you."

"Good to know, Kevin," Adam said. "You have no idea how glad I am to hear you say that. We'll bear that in mind. So anyway, are there any objections to the Tau Ceti idea? Or objections to continuing with this line of planning in general?" Though Adam said in his log that although there were quite a few very somber expressions on most of the faces in the room, no one suggested anything different. Mila's expression was reported as "simply beaming."

"This is an important moment," Adam concluded. "We'll adjourn for now since the ship needs our attention. I think we should meet again one week from now, and maybe every week thereafter if that's possible. Next time, I want to focus on the most important ship's system of all, the crew. We'll need to talk about things like our future inter-relationships with each other. By that, I mean what specific rules of conduct we should have amongst ourselves. You know, things like marriage, families, and children. These are the issues that will really matter if we are to refrain from simply destroying our mission from within. Again, this is all hypothetical, but paying heed to the realities of human interaction is at least as important as designing zero-G dishwashers. This is not the time to simply brag about what we've already accomplished. We'll really have to lay it out on the line here. We can't mess this up! I don't want to have come as far as we have just to turn this mission into some kind of "Mutiny on the Bounty" in space a few years down the road. Really, I've never been more serious about anything I've asked of you so far. The biggest threat to this mission of ours could be ourselves. Dr. Hosokawa, I think you'll be a major player here, not as a threat, of course, but as a major player in maintaining our emotional and psychological well-being. Mental health is a physician's field of responsibility, right? So I'll consider this to be your department. Any objections? OK good. We'll see everyone next week."

Later that day, Adam made rounds around the station while Mattie was on the bridge. Adam met with his engineering people at their various workstations. He arranged for a meeting with just them the following Monday. They were to discuss the developments in their department so far that he didn't yet want everyone to overhear. What he heard at this meeting was encouraging, although progress was not as far along as was originally hoped. Mangeet reported that his antiproton storage tank was in place at the far end of the now-completed platform that would anchor the various propulsion systems. Sparky the robot had been doing an exemplary job! Iza was in heaven! The various pipes and conduits for containment and direction of proton plasmas were being manufactured efficiently by the large 3D printer located just behind the radiation shield. Adam gave kudos aplenty to Mary Dupree. The new conduits could be attached and welded by Sparky as soon as they were off the printer. Next would be the magnetic field generators that would be arrayed along the long primary conduit. Trusses had also been extended above and below the lengthy plasma conduit to mount the magnetic chambers, which would concentrate and accelerate the Argon atoms that created the thrust from the ion engines. The Argon storage tanks would be manufactured soon and be placed between the radiation shield and the magnetic chambers. The Argon itself would be transported to orbit and pumped into the tanks when these were completed.

"So, can you guess what we're still lacking, Adam?" Azaan asked after his report.

"You know as well as I do," Adam laughed. "The main question right now would be how are we planning to provide all the electric power for all this? I also know the answer to my own question. When are you planning to complete the reactor shells?"

"The main reactor housings should be done two weeks from now," Azaan replied. "And believe it or not, we can smuggle up the control rods in some of our larger nuclear waste containers once the housing components are completed. We can say we jettisoned the cargos as usual, then brought the Shrike to the aft section to install them without anyone ever noticing. Sparky can install them unassisted. All the water and heat radiators will be in place by the time we're ready for the nuclear fuel. Not that I have any idea how we get that!"

"Let's discuss it, then," Adam said, looking at Mila, who was sitting just next to Mangeet.

"My contacts on the ground have been negotiating with the Pakistani Foreign Office," Mila said, rather softly, although it wasn't really necessary to lower her voice. They are now asking for four Shrikes and a basic array of support equipment in exchange."

"Four, eh? Not the three they originally asked for," Adam said. "Well, since we still have so many birds sitting idle all the time, I can't image Nathaniel will particularly object. They are paying a lot for them after all. So that means we will need nine overall."

"Nine?" Gregor gasped, but rather softly.

"Right," Adam replied. Four will be delivered to Pakistan permanently. One will be used to shuttle Mila from Uplift to Pakistan as an emissary and then bring her back here once she's done there, along with some cleverly disguised Plutonium. It's complicated, I know, but that's the only plan I've been able to come up with. You'll see."

"OK, so here's a bit more information," Mila continued. "We obviously will need to complete this operation just before we leave our orbit, not any sooner. Also, we'll need to orchestrate it so the station will be passing directly over Pakistan just as we transfer the fuel. Then we'll need to station keep directly over Pakistani airspace right until we leave orbit."

"Whoa, whoa!" Mattie said. "Adam, you all have apparently been thinking about this a lot longer than you've told me. This secrecy is a little unusual, thank heavens. Don't worry. I promise I won't let any of this leak out of my head or anything. I'll keep it a secret. But I still think I need the whole story here if it isn't too much trouble."

"There's only trouble if we screw it up, and then there'll be a whole lot of trouble." Adam began. "So, what we're talking about here is an exchange of four Shrikes and equipment in exchange for about 30 billion dollars for Uplift."

"That's kind of above the going price, isn't it? Mattie objected.

"Not if we include enough reactor grade plutonium to power up two nuclear reactors in the deal. We can smuggle that up to the station from Pakistan and leave before anyone can stop us. That should be enough nuclear fuel to supply electricity for our new station engines, and all the internal power for the station we'll ever need to boot."

"Well, that's pretty damned illegal, all right. I thought you promised Nathaniel you wouldn't break any laws!"

"It's not against Pakistani law to sell nuclear material to whomever they choose. Their government has already done so in the past. Lucky for us, they really, really want those Shrikes. Mila has been negotiating the deal. Remember, she's even volunteered to go down and escort those four Shrikes to their new home, and she'll bring an extra one along so she can come back up to us when she's done. She'll catch up with us directly over Pakistan. We'll time it so we're right over it too when she arrives with our Plutonium in her cargo bay. Once we recover her and the Plutonium, we burn ourselves out of orbit and out into deep space. No other nation's airspace has been violated, and no nuclear materials go into Earth orbit. So arguably, no laws will be broken."

"So how do we make this deep space burn without any nuclear material then, Mr. Smarty Spacesuitpants?" Mattie asked Adam.

"Well, we haven't quite figured out where to configure this, but I'll have Nathaniel send us up a conventional heavy booster, the heaviest we can spare, and have it ready to fire as soon as everything's aboard. The thing is, we have to be sure we're absolutely ready to go when we light that particular candle. We don't know just when that time will be yet, as far as I can tell. Mangeet, how goes the prospecting?"

"Prospecting, eh? I never thought of it that way. You know, if all I needed was a big sifting pan and some patience, I think that overall maybe I should have just stayed at home and been a prospector. Wouldn't work up here, though. As you no doubt may have noticed, my original projections have been a bit short in terms of anti-proton collection, but we are actually at about 75% of our original estimates at this point in time. It turns out that just because the computer thinks a deposit ought to be in a certain spot, that doesn't mean it will be. Imagine that! Then again, there's just as much of a possibility that certain areas might have more antiprotons than are projected. There's not much we can do that we aren't doing already. I'm sure we'll get to where we want to be in a reasonable amount of time. We still have a lot more construction to do anyway, not to mention all the required testing and other mandatory preparations. I'd still be willing to bet a ton of imaginary money that we can skedaddle out of here by May of this year."

"But that's just over three months from now!" Mattie said. "I know the sooner the better and all that, but I for one won't be holding my breath. There's a lot more prep work to be done, way more, especially for the crew. Not to mention the time we need to remember what we might have forgotten."

"Yes love, "Adam said. "I think we would all agree with you. I guess that all Manjeet's saying is that the ship itself could feasibly be ready in that time frame. We'll just have to play the rest by ear. Feel free to cool my heels for me if you see a need to."

"I haven't heard the answer to my original question." Mattie barked back. "You said we'd be needing nine Shrikes total. That's four for Pakistan and one to bring Mila up. What are those other four for?"

"For us, of course," Gregor piped in. "You don't think we're leaving Earth without a few Shrikes coming with us? What would we do if we found a place we want to settle? Or at least visit. Not to mention any uses we have no way to anticipate. We better bring them along, Mattie."

"Oh, yeah," Mattie said quietly. "Whatever was I thinking. You're right, of course. You own them all anyway, right, Adam? You can bring along as many as you want."

"Actually, four seems to be the best number, now that I think about it," Adam said. "We can't sneak away with too many more, though, what would Nathaniel do all day? These can start being delivered just as soon as I break the news to him. He'll just be thrilled, I'm sure. We'll have to have some of our people who can fly go down to Uplift and then bring them back up to here to stay with us. When we are finally are ready to do our little Pakistan escapade, whoever brings

Mila back up after she goes down to get the fuel will have to drop her off and then leave right away. Whew! This is getting pretty complicated! There's still a lot more to do before we're ever ready for this caper of ours."

Right on schedule, the next crew meeting came around the following Monday. Adam wrote in the ship's log that he remembered this particular gathering as the most consequential of the whole project. It seemed as much of a blueprint for ship operations as any actual set of engineering plans. This would lead to an actual written document that would govern all social interaction on the ship.

"OK crew!" Adam started semi-seriously. I hope you're all in a place in your duty assignments that you can socialize here for the rest of the day. This could be a long one." Mattie nodded in the affirmative on everyone's behalf. She had arranged everyone's assignments to allow for this. No one was absent. Adam continued.

"I plan to proceed as though everyone has already decided to stay aboard when we leave on our hypothetical 20-year-plus voyage to Tau Ceti. Assuming this is true, are there any thoughts about what each of you would need in order to most comfortably live your lives here? I know this is a bit awkward, to say the least. As a start, I think the main point I want to clarify, to be perfectly blunt, is whether you all might want to start families. And if so, how do you envision raising your children in this entirely novel environment? I assume you have all worked out that if there is an uneventful round trip there and back, all of us will be at least in our seventies when we finally return. If we stay at our destination for a time before coming back, we'll be even older. If we stay there permanently, our children could either stay there or have to return to Earth without us. These are very serious questions.

"And don't forget," Mila added, "that even if we only reach half-light speed, and that's questionable at this point, we will be aging at a reduced rate in comparison to the people on Earth we leave behind. It seems pretty likely that everyone who's there when we leave will have passed well before we get back."

Adam now made a confession. "You may or may not have noticed that I considered your backgrounds when I decided who to hire. I wanted folks with fewer ties to home than most other people. I kind of thought of it as though maybe we were all soldiers being sent to war. Who can know in advance who may or may not be coming back?" Adam was no longer mincing his words. "Does what I just said make anybody here less motivated to make a prospective journey like the one we are contemplating here?"

To Adam's genuine surprise, no one, not even newcomer Jeter, spoke out in the negative. Had he really chosen his crew that wisely? "Wow! That's actually not what I anticipated. I'm glad to hear it since I consider every one of you to be absolutely vital to carrying out this plan safely. I can't imagine having to take time to find anyone else to fill your places. We'd have to screen them properly and then take at least another year to train them once they signed on. Plus, since all of you have already given your all to turn our dreams into reality, I know you all feel you have a stake in its success. It's true. I wanted to commence this project while we're all still young, even if we will be aging more slowly." There was a hint of a laugh heard on the audio recording.

"We won't notice the difference ourselves," Mila interjected.

"Even so, it's time to cut to the chase," Adam continued. "Now, if I use my calculator, I notice that there are currently eight males and nine females on the crew at this moment. I guess I need to know if we need to pop back to Earth and abduct a man for one of you ladies to keep you company. What I guess I'm really asking is, who here might want to procreate on a spacecraft that won't see Earth for decades?"

Only two failed to raise their hands. These were Kevin St. Patrick and Victoria Zasco, the pharmacist and the dentist. "OK, so what are your objections, the two of you?"

Victoria responded merely that she never really liked the idea of bearing children. She simply couldn't wrap her head around the idea of either being pregnant or going through the physical process of giving birth. She had always either believed that she would never marry, or at least even if she did, then she would prefer to adopt a child when she decided she was ready. Kevin stated flatly that, firstly, he didn't believe the ship would ever manage to leave orbit, and secondly, he didn't like the idea of just arbitrarily being told to marry someone at this meeting. "Alright! Adam sort of cheered, but not in any way that sounded angry or disappointed. "This is the sort of honesty and frankness that we have to have if we want to coexist together." Now, Kevin, I guess we will need to assume that you will commit to our mission here if we are able to move our theories forward. Would you be happy staying with the crew without committing to raising or even just making children?"

"Well, this is getting a bit out there, isn't it?" Kevin asked.

"Now's the time to decide, out there or not. Personally, I just feel in my bones that we will have to carry our trip into another generation. With a limited crew, and it will have to be limited, I can't but feel that we'll need to stretch the boundaries of the family norms that we have on Earth. We should and will have marriages if a marriage is desired. Not that all marriages are normal down there these days. But three couples here are already legally married, and we seem happy enough. None of these marriages have yet produced any children. In the case of Mattie and I, this is completely intentional. Dr. Hosokawa has been helping us with a birth control program since he came on at Uplift. Dr. Annison has taken over with this since she came aboard." Juni and Jia both confirmed that they had done the same. "I'm feeling pretty sure that when we feel the time is right, we'll give parenting a shot. That doesn't mean this will be a requirement or anything. It's not like I'm trying to build up some sort of sex cult here or something. It's just that I personally don't want to have our ship go on automatic if the mission outlives all of us. Maintaining a human presence is kind of the point of the whole mission. We could program the ship to document what we find out there, and transmit it all back to Earth, but that's not what exploration is really about."

"On the other hand, why don't we just do that now?' Kevin ventured. "We can all go back home when the ship is ready and just send the station on its merry way."

"Is that really what you think this is all about?" Adam asked. "Think what would have happened if someone like Magellan, or whoever, just wrote a message and put it in a bottle and threw it in the sea." If anyone's out there, just write a note about your life and throw the bottle back. We'll figure you out eventually." There was again just a tiny bit of laughter heard on the recording. "Nope, as far as I can tell, humans have to see things for themselves if they want to make them real. Think of all the pictures taken of the moon and Mars before anyone ever went there. We knew we'd get there eventually, but we still felt the need to go. I just don't want to wait for eventually. I want to actually travel to another star! If this ship can do it now, then I'm going, even if it's just me and the bots."

"Hey!" Mattie shouted.

"Well, I'm pretty sure a few more of you might be coming," Adam said. "Think again for a moment how all those explorers took fully crewed ships, even if no one on them had any idea whether or not they would ever come back to their homes again. A significant number of them never did. Be it death from storms or war or starvation, they still signed on to risk it all in spite of knowing the hazards at the start. That's the only reason they found out they had a larger world."

"Sorry to say it, but those explorers really were cult leaders, weren't they?" Kevin responded. "If nothing else, they were certainly religious fanatics. Even worse, they assigned themselves the God-like powers of life or death. They'd whip you till your skin fell off for preposterously trivial reasons. How do we know all this won't go to your head, too?" Kevin really had become brutally honest.

"Well, that's what we need to dish out." Adam agreed. "Maybe we should come up with some written agreement. Something like, say, the Mayflower Compact that was drawn up off of Plymouth Rock. Those poor souls had already fled from religious persecution and desperately wanted to at least avoid tyranny from within their own ranks. I think we should draw up something like that before we ever really commit to leave. We aren't in jeopardy of tyranny from the ground, unless they deduce what we have in mind somehow. Our greatest threat might actually be ourselves if we don't stick to an enforceable code of conduct. I guess maybe I might wind up being a cult leader after all, who knows? But a set of rules would be the best way to keep that from happening. If this is going to work, we will all need to come up with these rules together. Americans like democratic principles in their society anyway. No system is perfect, but democracy is the best strategy the world appears to offer. Anyway, this is all just food for thought. Let's get back to what we were talking about before. Kevin, this is not a reprimand or anything. For goodness' sake, keep up with the skepticism. Let's get back to your original point. You certainly have the right to refrain from marriage or procreation if you so choose. I have to say I don't think we have time to have you go back to Earth to find yourself, someone to run off to our star with. I'm afraid the woman, or man of your dreams, is already at this table waiting for you."

"You just don't want me reporting to my Space Force people about what's going on up here. And no, I don't happen to prefer men to women."

"As long as we're being frank," Adam said, "You're right. I guess I don't really want you blabbing about our plans to the Space Force or anybody else. For one thing, we still have no way of knowing whether or not we'll ever be going anywhere other than around and around our current silly rock. If you go casting aspersions to the people in authority, they probably won't let us even conduct any research, let alone try interstellar travel. Secondly, what gender prefer in your significant others is no damned business of mine. It only matters who I prefer, and that would be Mattie."

"OK, we're really digressing here," Peter Hosokawa interrupted. "I thought you said all this was going to be my purview anyway, Adam.

"That is a very good point," Adam agreed. "Especially since the topic is procreation. Please take over, thanks."

"From what I have studied on the topic," Peter began," I feel sure that in instances of isolation from one's usual social settings, humans generally need to resort to unusual habits merely to survive. For example, Victoria. I think if you don't wish to marry or even mate and have children, there's no need. But I think it's inevitable that you'll need to participate in raising whatever children may come along in our small group. We'll need to have people teaching school for example, or babysitting if the parents need to have some alone time. But I don't think anybody expects you to just breed for the sake of it." Victoria gave a very quiet statement of agreement with this. "On the other extreme," Peter said, "there is such a thing as artificial insemination." There was a sort of mildly suppressed murmuring heard here. "No really, a surprisingly large segment of the population down there below us that was conceived in this way. I'm just saying that if you ever do change your mind, and the urge overtakes you, you can be a single parent here on board. I don't imagine anyone here would object to that."

"Well, I do have one objection!" Victoria responded. "Since I don't think we can order take out from up here, that means that at least one of you will need to be the donor. Am I wrong?"

"Well, I haven't asked any of you fellows who might be willing…"

"We're all able, though, I feel sure…" Gregor piped in, with a response of hardy laughter.

"I imagine all of us could contribute to the cause," Adam said, and not just for one pregnancy or one particular mother. You don't even have to know which of us it is if you don't want to. Besides, who knows what situations may arise? One of us men may become infertile, or all of us. It might be good to have, shall we say, a backup plan."

"For God's sake, Adam," Sharon said. Why do you Engineers always have to be so clinical? We're human beings here, not some subsystem you have to operate!"

"True enough," Adam admitted. "But as we've been saying, people just do what they have to do sometimes. It's plainly a matter of survival. Anyway, fulfilling the need for love and marriage is one of humanity's greatest challenges. I'm hoping we don't all get bogged down with things like jealousy and infidelity as time goes by. Those are the things that destroy human society every day of the week. Even if this does make me a cult leader, I think we will need to show true flexibility in how we deal with these issues over the years. We can't afford not to forgive each other when we screw up. We may well find it necessary to be, shall we say, unconventional in our relationships."

"Well, at least I can confirm that no venereal diseases have been brought aboard the ship," Sandra volunteered. The response to this was a minute or so of complete silence.

"Well, yes, that is good to know, Doctor," Adam said rather slowly. "We must all admit that we are, in fact, young adults who will be locked up for life in an enclosed space. Again, I think in the near future, we should codify our conduct. A good deal of compassion for each other will be necessary. Vital, in fact. Logically speaking, I'm expecting that most of us will continue to pair off. Again, I am, for the moment at least, the captain of a vessel possibly soon to be underway on a very long cruise. International law authorizes me to perform legal weddings. I imagine we can pull up the required forms?" Juni nodded in the affirmative. Nevertheless, if anyone decides to pair off, or do more than pair off, or not pair off at all, that is no concern of anyone else's. We will simply do what we need to do to stay happy. Infidelity won't be a crime, but it may cause discord. I'd just ask that you get permission from your partner or partners before you do it, although that's not something I'd like to codify. There, you see. Things can always get more complicated, and they're complicated enough already. These kinds of behaviors will only be discouraged if they cause discord among us. Just remember that when the time comes to really commit to continuing with this project. That's all, I think." The meeting was dismissed at this time, to the probable relief of all its participants.

The final meeting of March 2088 was dedicated to trying to think of tasks that were believed to be essential to accomplish before actually seriously thinking about leaving the Earth. There were a few items that had been discussed earlier which were not yet accomplished, and some things that hadn't actually been considered before. In the first instance, the communication satellites, or relay stations, really, since they wouldn't be orbiting anything. When asked, Juni reported that these had actually been designed, and an order could be placed for twelve of them to be manufactured. She hadn't placed an order yet because no one had told her whether or not any nuclear batteries had been purchased or, if so, when they might be available. Adam directed her to go ahead and purchase the satellites. "Can you manufacture some sort of catapult or launcher in the aft section that we could use, Gregor?" she asked. Remember that if we simply jettison them, they'll just keep running parallel with us the whole rest of the trip."

"Oy, that's right," Gregor said. We'll have to use maybe a series of the thrusters we use to get rid of our space junk. Even that won't begin to slow them down enough, but it's better than nothing. Wait! How about adding a set of sails that can catch our thrust energy trail? They'd have to be pretty resilient to stand up to that, but I'll look at it right away."

"Please do," Adam said." If I thought we couldn't send any messages back home whatsoever, I'm not sure I'd want to leave in the first place. It'd be just like we never existed, as far as the world was concerned."

"We'll make it work, Adam," Juni said. "The computers say we can generate a signal of sufficient integrity as long as we can slow the relays down enough."

"If you're not able to, please let me know as soon as you know," Adam said.

"That's a promise," Juni confirmed.

"Anybody else?" Adam asked.

"Yeah, I've got something!" Jeter spoke up. As the newest crew member, he hadn't really had a chance to contribute anything to these meetings before. Adam wondered if his input might inadvertently spill a bean or two about the under the table nuclear plans. But his quest actually had more to do with astronomy. "My understanding is that there is long-range radar already placed on board?"

"Correct," Adam answered.

"I'm assuming this is meant to detect any unexpected objects in our path during deep space flight. I wonder what we plan to do if there are any?"

"On all of our interplanetary probe flights, there has never been a consequential impact with even the smallest micrometeorite. There are no such objects thought to be present in interstellar space. But we should still be cautious. If we detect any ahead of us, we should be able to maneuver around them in plenty of time." Mila made this pronouncement with the tone of authority.

"True enough," Jeter said. "But we're going to be traveling at speeds orders of magnitude greater than anything that's been experienced before. I'm aware that even if we bring some sort of defensive missiles or something, there is not much chance of them being effective. Besides, as Adam has been saying, we never know just what sorts of threats we might be facing."

"What, like aliens or something?" Manjeet said with a semi-amused sort of tone.

"Well, that's pretty unlikely to be the threat," Adam interjected. "But, you know, I suppose one never can tell. I guess maybe a missile array of some sort is worth looking into. We can't really tell the ground about it, though; they'll think we've gone commando or something and want to attack the Earth. If anyone has any ideas, let me know sometime later. I guess we could mount a few at the front of the ship. We'll put the bots to work. We could control them from the bridge, right?" Gregor nodded in the affirmative. "Anything else?"

Manjeet cleared his throat, indicating he wanted to be called on. Adam did so.

"I imagine I should report that we could conceivably be prepared to perhaps do some engine testing sometime in May, around a month from now." There was a bit of surprise, not actual shock heard in the background on the audio recording. "No, nothing spectacular, just routine stress testing, if we all agree to proceed." Even Adam was not expecting things to have progressed to this stage so soon. He certainly wasn't expecting Azaan to announce it if it had. He strained to maintain an air of detachment. "Soo, do we have enough fuel on board to do this?" Without saying it out loud, he was asking Azaan if he was implying that a sufficient number of antiprotons were now collected in the storage tank. Those in the know understood it, too.

"Yes," Azaan replied without twitching a single facial muscle. "We received the last shipment two days ago. It's all squared away."

"Phenomenal," Adam light-headedly replied. "Then we can actually start a bit of testing, eh?" He sounded rather shocked, as a man does when a distant dream actually seems like it has a remote possibility of coming true. "You know, just for kicks, I just remembered an unfinished bit of business I've been putting off a while. This seems like a good time to bring it up."

"And that is?" Mattie asked.

"Well, the Uplift Orbiting Research Facility seems a pretty cumbersome name for a ship that may one day be an actually interstellar spacecraft. What is the acronym? UORF? That doesn't seem like it would do, does it? It's not even pronounceable! When we meet next week, let's have a little naming contest. We'll vote on it. The winner gets a salmon dinner. Is that OK, Sharon?"

"Sure, we've been able to breed some. Kind of tricky too, you need to go from salt to fresh water at just the right time. Anyway, we'll have plenty of salmon from now on."

"Yeah, I'll bet you already have a ship's name in mind! And you love salmon. This is going to be pretty unfair," Mattie said with her usual tone of sarcasm.

"It doesn't matter who wins, even if it's the one who furnishes the prize," Adam said. "I think we'll all be requesting a salmon dinner soon, though, now that we know we can. OK! Anything lacking on the medical side?"

"For those of you who haven't donated blood yet, I'm still waiting!" Kevin responded. "Adam! Is it true you're really an AB negative? I'm not buying it. I think you're just desperate for attention."

"Nah, it's true. Don't fret. It's not like I requested it before birth or anything. It doesn't imbue me with mystical powers either. It's not like I had it as a goal to be different from nearly every other member of my species. Anyway, OK, I'll stop by this afternoon. You know I can accept any blood type that's rH negative? I can only donate to other AB recipients too, remember."

"Gee, I hadn't heard," Kevin said sarcastically. "You're the one who said everyone had to donate, as I recall. I guess you might need it yourself I suppose. Don't worry, my massive needles are sterilized and reasonably sharp. No excuses, please."

"Gotcha," Adam said simply. "Anybody else?"

"Oh, yes," Xao Jia raised her hand. "I'm afraid I'm really not satisfied with the ultrasound machine I requested. I guess I should have been more specific about the brand. I'm really more comfortable with the Chinese-made models. I'd really hate to not have the one I prefer. In my profession, and with nearly mandated pregnancies on this mission, I think I really should have one."

"Get with me after the meeting," Mattie said. "I'll send down an invoice."

"Excellent," Xao Jia said. "I think that's all I still need."

With that pronouncement, Adam declared the meeting adjourned for the week. He gave a subtle nod to the Engineers to remind them that tomorrow's little subcommittee meeting would be held as usual. Adam sensed things were truly picking up speed.

Chapter 15
Countdown to a Countdown

From transmitted meeting minutes and ship's logs...

Adam fondly remembered the next day's engineering meeting in his later dispatches. At this gathering, he asked for a status report on all propulsion systems and outstanding project requirements. Azaan confirmed that earlier that week, the collectors happened on a "motherlode" of antiprotons over Norway in a magnetic void pocket of the Van Allen belt. A full two grams had been collected, exceeding the 50-gram total required by an entire 75 grams. These particles were already locked away in the containment tank and were now swimming around quite contentedly. He reported that the liquid hydrogen tanks were full, the magnetic field generators were operating as expected, and the feed system appeared to be functioning as exactly designed. Azaan was, justifiably, especially proud of this. He had even decided to transmit down a patent application from Earth orbit for the first time in history. He had bypassed Uplift control when he did so, however, in the hope that no one at the company would get wind of what was happening up on the station. His design called for antiprotons to be fed into the main engine shaft just where it met the engine bell, or nozzle in the more common vernacular. The trick was to feed the antiprotons into a stream of accelerated hydrogen ions, well, protons, one at a time. The frequency of feed was to be one antiproton, either every one or every two seconds, depending on just how fast you wanted to go. The plasma stream sending the protons down the tube would propel the resulting explosion out past the bell before the explosion could destroy it, thus pushing the ship forward with rather extreme force. This strategy almost seemed counterintuitive. Common sense would make you think the resulting explosions would simply blow the ship or anything else in the vicinity to atoms. But the borophene composite material used for all this plumbing was designed using supercomputing technology, and every simulation indicated it would work. Conventional rockets used combustible fuels to produce controlled explosions as well. The problem was that massive amounts of heavy fuel were needed with rockets for results that lasted a very few minutes at most. For Azaan's new engine, five grams of antiprotons would combine with five grams of ordinary protons to send the ship to a nearby star within about 20 years. Once they arrived, if the crew so chose, they could then turn around and come back to where they started. The nerve-wracking part was that these antimatter reactions had never been accomplished before anywhere except perhaps in a huge supercollider in a remote underground lab.

One bonus about being in space was that it was a really super cooled environment. All the engine components would also have supercooled hydrogen, cooled down to a slush. Actually, that was still fluid enough to flow through the various components and then through an intricate and quite extensive net of radiators. Theoretically, this would keep the whole engine from simply burning into space ash. Also, theoretically, such a reaction would possibly obliterate the ship in an instant. But if it didn't, they would soon be speeding through space at potentially half the speed of light within a few short years. It would also be extremely difficult to slow down and return to earth within a time frame under a few more years. They all knew that this idea was as close to being a suicide pact as it could be without actually pulling a trigger. The people at this meeting were now committing to loading the pistol. Nevertheless, no one there apparently had even the slightest apprehensions about moving forward. Juni stood by her simulations. Mary Dupree trusted her materials and manufacturing. Iza was really proud of Sparky. No one on that bridge had any doubts, whatever. Mattie had devised a test protocol where regular protons would be fed through the system without magnetizing the field generators to verify flow rates. Of course, nothing but a simulation could ever be done without the two nuclear reactors and generators up and running. And that couldn't be accomplished unless Mila and her Pakistani pipeline were successful in obtaining some Plutonium. Manjeet pointed out that at least the fuel components were aboard and ready. After all, only 5 grams of ionized hydrogen were needed to counterbalance the 5 grams of antiprotons needed for each leg of the trip. There were currently over 20 tons of liquid hydrogen currently in storage tanks throughout the ship. That didn't count the enormous amount of liquid water aboard that could be broken done into oxygen and hydrogen using the nearly endless supply of electricity the nuclear reactors could supply. They realized that they actually would be quite fat on resources for this trip. Adam truly thought he was only dreaming that day.

Gregor and Manjeet reported again that the two reactor casings were completed, with control rods in place and cooling liquid ready to go as well. This dedicated supply of liquid hydrogen coolant was stored in slightly heated tanks to prevent it from freezing out there in the freezing space environment. These tanks were also located at the back of the engine frame, just waiting for the Plutonium to arrive. The solar panels were providing plenty of electricity for the time being. Gregor also reported that the solar sails were ready to be installed by Blackbeard, Popeye, and Sinbad whenever the order was given. They would be attached to the back of the large solar panels themselves, making for a rather voluminous but lightweight array. Gregor was successful in designing these very thin sails to enable them to also collect any hydrogen atoms the ship encountered as it moved through seemingly empty space. They would then be shunted into a fine net of tubing woven into the sails and then pumped into collection tanks, although it would be a very slow process. The sails were not originally to be installed until the ship left orbit to avoid causing suspicion, but Adam decided that since the station was purported to be conducting research, it might serve a better purpose to deploy them in advance. Besides, there was no real chance of damaging them by accelerating rapidly through a vacuum. At least they could verify that the sails were working before they throttled up to full velocity. Another contingency they discussed was what to do if things simply didn't work as intended. It had been previously agreed that the best place to first light the antimatter engine was not in Earth orbit. To actually begin their departure then, it was decided that not one, but two big conventional rocket engines full of fuel would be lifted into orbit as early as next week. These hopefully would take them far enough away from the planet to distance it from any fatal malfunctions when the antimatter engine was activated. Manjeet confirmed that mountings for these two monster rockets were being added to the rear truss system as they spoke.

A formal plan evolved to actually execute this momentous departure. It was decided that just prior to committing to leave orbit, a test of the station's maneuvering system would be performed. The ground would be informed of this in advance, as it would certainly be noticed. Next, Uplift would be notified that one of the large rockets would be fired for the purpose of testing the station's tolerance to stresses incurred with acceleration. Of course, everyone on board already knew that computer simulations confirmed that the tolerances were, in fact, far in excess of what a mere combustion rocket would produce. At the appropriate time, it would then be announced that the station was being taken on a short test flight. The ground ought not to be concerned about such a routine-sounding dispatch in the slightest. Such test procedures had been a routine evaluation requirement since the days of Chuck Yeager and Scott Crossfield. The ground would be told that after a run of a few days, the ship would be turned around

180 degrees, then other rockets would fire, and the station would return to Earth orbit. In actuality, though, there would soon be an unexpected surprise for the ground controllers. Assuming all systems were nominal, the Earth would be informed that perhaps it wouldn't be a bad idea to keep sailing outward for a while. Updates would be transmitted regarding when they would actually decide to return. Perhaps a bit later, the crew would announce that everyone down there might just need to be patient for a few decades.

Mila made the point that since touring the solar system had been ruled out in favor of a direct interstellar course, the ship would have to maneuver in ways that would not support the excuses they were making to Uplift control. The direction of departure would be outward and to the south, about 30 degrees shy of the polar axis of the Earth. It would make it look like a huge navigational error had been made, if not a major malfunction of some kind. It was then decided to go ahead and tell the ground at that point that there had been an error with the navigation computer and maybe other equipment failures to boot. The message to Earth would say that the crew was safe and that it would require time to fix the problems before they could return sometime in the future. Adam thought these ideas sounded better than his plan. They would probably wind up revising their ideas as they went along anyway. He also said he wasn't particularly worried that the Earth would somehow try to intervene. That wouldn't really be in their power to do. Once the engines were lit, they were on their way. It would be very nearly impossible to stop them from leaving at that point on. By the developing checklist, when the computers thought it was safe, the solar sails would be deployed, and the ion drive would be activated. All systems would then be evaluated by the crew. If all was verified and then reverified to be safe, the antimatter engine would then be ignited. Only after they were well underway would the Earth be informed that they would not be returning. As to other pressing concerns, Adam next took the opportunity, since interactions with Pakistan were already the topic, to ask Mila to communicate with her contacts to see if a dozen or so surface-to-air missiles might possibly be obtained from that government as well. These would need to be of a type that was capable of operating in the vacuum of space. Gregor reported that he was already working with the bots on constructing a launch bay dedicated to the missile array. This was located in one of the forward wheel cargo spaces. A new hatch was being installed that would open on command, and the rocket platform would then extend outside before launching. In that way, all of the necessary controls, wiring, and equipment would remain shielded from the elements, or rather lack of elements, until such time as the weapons might actually be used. They were to be kept in an unarmed state until that time, if that time ever actually came. Mila promised to follow up with all of these complex and clandestine arrangements in a timely manner. She also rather gleefully reported that her astrophysics lab was now nearly all up and running. All the sensor equipment was functional, and she had already begun to start every workday taking readings of surrounding radiation levels, temperature, and other required data. Interestingly, the Tau Ceti star was more visible in the southern hemisphere than it was in its current northern orbit, so it was not fully visible as frequently as she would have liked. But then again, it was in every way crucial to observe it as closely as possible in order to accomplish the mission they were planning. She was now in the process of programming her telescope to make frequent sightings for future navigational purposes. She was nervously conscious of the huge importance of her upcoming assignments and the possible international controversies they would soon be causing. This would be true whether she was successful in accomplishing them or not.

Daily activities continued at the station as usual despite everything else that was happening behind the scenes. As always, confirming the feasibility of adequate food production was paramount. Sharon confirmed that the general fish population was increasing at a rate pretty much approximating what it would be if the aquariums were on Earth. Sharon concluded that it would be more of a problem to contain any population explosions over the coming years than it would be to run out of fish. She confirmed that there would be plenty of room for freezing and storing both the fish harvests and the projected crop harvests as well. Otherwise, any surpluses would have to be jettisoned overboard. The idea of schools of frozen fish being added to the fabric of the universe apparently was somewhat disquieting to her. Another factor that she had worried about, and which now appeared to be a truly impactful one, was the absence of pollinators aboard the ship. During the planning phase of the mission, no one had wanted to hear even the suggestion of introducing any sort of insects into the enclosed spaces they occupied. Accordingly, great lengths were taken to screen Shrike cargo bays before any payloads were brought aboard to avoid introducing unwanted multilegged passengers. The idea of ever swatting flies, squishing cockroaches, or, most especially, running from swarms of bees on the station was unthinkable to the crew. Additionally, accidentally introducing them into some otherworldly environment didn't seem like a good idea either. The equally important flip side to this argument was the absolute need for the plants on board to exchange pollen in order to produce anything edible. Pollination was absolutely needed to ensure the continuation of the production of seeds for next year's crop. Therefore, duties for whoever was working on farming tasks necessitated education about how to pollinate the numerous species of plants manually. In order to alleviate the need for millions of cotton-tipped applicators over a period of years, the crew was taught to pick one flower at a time and swipe it over the appropriate parts of the other flowers. This massive vegetative orgy would be immensely time-consuming, though. This process was nowhere near as efficient as onboard bees would have been, but it did seem to work well enough. It was also a process that would need to be done year after year for the entire duration of the mission, even if a new bee-free planet were eventually colonized. Fortunately, the various trees aboard seemed able to pollinate themselves with only the air currents present in the arboretum. This, at least, would be extremely useful once the trees grew to their adult sizes. It was realized that if gravitational forces were introduced with forward acceleration, such air currents could occur. There thankfully, also was a large supply of seeds for these larger species brought on board as well. The seeds were projected to be able to survive the trip and then propagated on whatever habitable areas might be found at the end. Future forests might even result, or so it was hoped. All in all, it became apparent to the crew that no starvation scenarios appeared likely to happen aboard their ship, no matter how long they were away from Earth. This was a great positive influence on morale, and that was definitely not a bad thing. No one would want to go anywhere without an iron-clad food supply.

The next week's crew meeting was a positive experience yet again. Adam began as he had promised with the preplanned discussion of possible names for the station. It seemed important to make their domicile feel more like a means of transportation than like a mere floating container full of people. Suggestions included short lists of the usual heroic sounding ships of history. Constellation, Constitution, Challenger, Columbia, Intrepid, and so forth were all put forth with the raising of hands. These were duly put up on the flat screen for posterity. Then, knowing that it was probably expected of her, Mattie called on Adam to express his view on the matter. Also as expected, he embarked on a small diatribe explaining why his name was the best choice.

"Now you know," he said, "that this mission we're considering will be far and away the most audacious, unprecedented, heroic, and undoubtedly foolhardy journey anyone has ever attempted." (Of course, they knew)." To me, though, we resemble more a group of settlers making a desperate journey over desolate territory with very few resources. I believe every nation on Earth has some similar sort of story somewhere in its folklore." (Here we go…) "From the Vikings to the first polar explorers to the Moon and to Mars, humans have this odd habit of risking their lives to set eyes on new horizons. I have no way to rationalize or explain this. Objectively, it seems kind of insane. But exploration always has and always will be done by human beings. If we don't do it, someone else will. (Yes, we know…) Now in recent American history, these wandering people were simply called settlers. They were generally poor and disenfranchised people who had to give up everything they had and decided to move west. I'll be the first to say that there were as many evil consequences accompanying this movement as there were eventual benefits. But simple human nature seemed to be its major driving force. I'll tell you this, I'll be the first one to advocate for doing all we can to not export any of our many human foibles to wherever it is we may be going. But as always, there may be unintended negative consequences anyway. We should all be on our guard. Oh, now where was I? Oh yeah, as far as my preference for a name for our ship, I say let's pay homage to that multitude of travelers common throughout our history. This is who we hope to be as well. How did they do it? American pioneers most often used just a simple little covered wagon. They had to pack as wisely as they could to make their nearly impossible journeys. Even so, it cost them their life's savings to finance. On top of that, they also could foresee that their travels could easily cause their own deaths as well as the lives of any children or family they brought with them. But they traveled nonetheless, off to seek their dreams. Our little wagon here will one day also be deemed as primitive as theirs were in their time. Those wagons were called Conestogas, after the company that built them. No matter what we decide on, we'll always be aboard a Conestoga in my mind. That's my choice to name the ship, the Conestoga. I know it's a bit corny, but I love it.

A few hmphs and hmms were recorded, but no one rose in disagreement. No further nominations were made. Adam called for a secret ballot. This called for pens and pieces of paper, which were, in fact, present on the bridge. These were passed around. At this point, there was still a supply of these writing utensils present on the ship. When these had been used up, they would go totally paperless unless someone figured out how to make more. No one had remembered to order another supply. After a few moments to count the ballots, to Adam's great delight his preferred name won the contest by a wide margin. Other candidates included one "Enterprise" and one "Barack Obama". But in retrospect, it seems inconceivable that anything other than "Conestoga" could ever have been chosen.

Adam then announced that it had been arranged to assign four Shrikes to the ship on a permanent basis beginning that week. He first asked if there were any preferences as to which particular ships should be assigned. All crew members had varying experiences flying aboard company spacecraft, and he thought that certain of them might have been favorites of the crew. The numbers of any ships that were mentioned were taken down. Adam would try his best to get the four most desired ones. And just like the ship itself, he then suggested that those Shrikes be individually named. As before, he already had decided in advance the names that he himself preferred.

"I know all of you are pretty familiar with space exploration history," he began his second diatribe. "You must be, mainly because I spend so much time lecturing you about it. Now, lots of spacecraft have been named for astronauts from the US and elsewhere, and justifiably so. I'm really a history addict, as you all know! One thing that always bothered me is that there were a few of those very first NASA astronauts that everyone seems to have forgotten about. It's a bit of a pity really, for various reasons. Unfortunately, we have only been able to get approval for four Shrikes to claim for ourselves. But there were actually five men chosen early on for the program who were killed during astronaut training. Anybody remember their names?" Adam was not really surprised when no one could name a single one of them. "OK, let's see if I can get them in order…I might not. I believe Ted Freeman was first. He sucked a goose into his T-38 engine on Halloween 1964, near Ellington Air Force base. The subsequent crash was fatal. I remember that because I got all this information from Mike Collins' autobiography. Mike's birthday happened to be Halloween, as well. Next were Elliott See and Charlie Bassett. They were sharing a T-38 flying to the McDonell Aircraft Building in St. Louis, I think. They were the primary crew in training for Gemini 9. It was put down as pilot error, but they actually crashed into the very building they were planning to visit. I'm not sure how or why. Then there was Ed Givens. He accidentally ran his Volkswagen off the road after going to some flyer's fraternity meeting in Houston. I think that was June of 1967. Finally, there was C.C. Williams. Some sort of aileron failure on his plane near Tallahassee in October of '67 led to a fatal crash."

"Sure, naming our Shrikes after them sounds fine to me," Mattie said quickly. "But five guys for four ships feels like we're one Shrike short. Could we combine two for one of them, like the Bassett-See or something?"

"Yeah, I don't know how to make it work either," Adam said. "I was sort of thinking that if we ever make landfall someplace, I assume we'll be 3D printing land vehicles of some sort, like a rover or something I guess.. Maybe we can call them Givenses, like a trademark."

"Not good enough," Sandra said. "Maybe if we ever name our first settlement, if there is one, it could be named for him."

"You know, on second thought," Adam said, shaking his head, "There's nothing at all wrong with the "Basset-See." Those two names are already linked together historically. Let's make them known in another solar system as well. We'll decide what to name a colony whenever we actually build one. Humans value names. Why else would every sailing vessel ever built always get christened? So, any objections to my last suggestion?" There were none, so a motion to adopt was made and passed, then recorded in the minutes.

With nothing left to discuss, Adam chose this meeting as the time to make what he considered to be his most significant announcement to date. He had no doubt been dreading it, but he knew it would have to be made eventually. Why he chose this day was unknown, but for some reason it seemed a good day to him. It came out a bit awkwardly.

"OK then, I guess this is as good a time as any to tell you. Please just try to hear me out. It appears from all our testing and simulations that it may, in fact, be possible with the technology we now have available on this station, this Conestoga, to, in fact, leave Earth orbit and travel to whatever destination we choose. So far, the majority of us here, all of us in fact, have specified their preference to travel to and explore the recently discovered exoplanets at Tau Ceti. What we have yet to vote on is whether or not we should, at this moment, actually commit ourselves to commencing this mission. As I said, we have in place the necessary hardware and, um, software," nodding to

Juni, "to attempt this. What I think we need to still verify is the current preparedness of our most vital component. That component being ourselves. If we humans don't function with the same level of reliability and dedication to our mission as the hardware does, then we should just get off the ship and shoot her off on autopilot. I'm thinking that maybe the best way to ascertain our own readiness is to get together one week from now and codify what behaviors will be acceptable and correct. Pardon my incessant analogies, but this situation really makes me think of the Mayflower. That storied ship was also filled with pioneers. As everyone knows, they codified what they considered their obligations to each other when they drew up their Mayflower Compact. It was really pretty simple. They merely wrote that they all swore to help each other live harmoniously to meet their goals to the best of their various abilities. That was it. I think our Compact should be a bit more comprehensive, though. Not obsessively so or anything. Maybe it should be more like a kind of constitution that defines who will have authority at any given time. It should also be a tool that confines this authority. There's no way a dictatorship type of situation would work in our particular case. Most of us here grew up in a democratic system. Maybe our arrangements can be loosely based on the criminal codes that modern naval vessels use, with modern democratic morality. We'll have to look at it and see what we can work out from there. Anyway, my point is let's see if next week we can draft some sort of Conestoga Compact that we can all stick to. I think that if we can manage to draft something along those lines, then we might reasonably feel like we can commit this ship of ours to some sort of voyage. Is that acceptable?" Peter stated out loud that this all sounded like some sort of conditioning exercise for what once again sounded like a cult initiation. Kevin St. Patrick, of all people, countered him by asking if there was some other way of persuading the crew to fully commit to the mission. Peter conceded that there was no better plan that he could personally think of. But the bottom line was that minutes clearly read "Unanimous affirmative nods" on the question of a Compact. Apparently, Adam had succeeded in persuading them that such an agreement was needed.

Another week passed, and the next crew meeting commenced. During the week that just passed, the ground crews observing the activity on the Uplift Orbiting Research Facility couldn't help but notice an uptick in the number of Shrike missions dedicated to servicing the station. Adam and the engineers knew that any ground observer might realize that there was something unusual brewing up at the station. From a new set of robot surgeons to a new ultrasound machine, to several new auxiliary trusses and two new liquid hydrogen tanks, not to mention even more liquid hydrogen, it struck several high-level staffers, including Nathaniel Floatingfeather, as more than a little bit odd. Such upticks always made him a bit apprehensive, given what he and Adam had discussed with his last visit to the station. When Adam invoked his authority as director to have two fully fueled Ultra-Lift liquid rocket engines ready to send to the station within the next 1-2 weeks, he felt sure something unusual was up. Adam didn't just want those engines fired into orbit; he wanted each one fully loaded with fuel and placed on top of another one. The lower one would be used to lift another fully filled booster rocket to rendezvous with the station. The lower stages would by then have exhausted their own fuel supply and would then be jettisoned to burn up in the atmosphere. In point of fact, nothing quite like this had ever been done before. This was also obviously something to be concerned about. Nathaniel, being the loyal chief of operations that he was, knew there was not an awful lot he could do about a direct order from the owner of the company, so he felt obliged to comply with Adam's request. In any case, Adam was the legal owner of all the company's hardware, so what he said counted. Nathaniel, therefore, braced himself to hear some sort of unusual news in the near future. As the station again became a beehive of activity (that had no bees on board), another week passed. It was soon again time for what would probably be a fateful meeting of the crew. This one would have its share of interpersonal contention, and it kept everyone away from all of their usual afternoon routines. But as Adam had warned, it turned out to be immeasurably important. The document they composed was destined to be one of lasting historical interest.

The meeting began routinely enough. "Welcome everybody," Adam commenced. "To start, I'll just say that nothing has arisen this week to contradict or postpone anything that was said at our last crew meeting. What we're discussing here today is still merely conjecture. We'll start with the premise that in the next several weeks, the Conestoga, our little home here, may be capable of launching itself on a journey that may take decades or even our entire lifespans and beyond. If we ultimately agree to do this, I believe our crew must lay down and then comply with a set of basic principles designed to ensure that our lives will be as peaceful and tranquil as we can make them. Any distress or lack of harmony amongst us would, therefore, only arise from factors other than crew relationships. It seems to me that over the past year, we have succeeded in manufacturing for ourselves a small village in which to live. It's only unusual in the fact that it can move from place to place. We have all the food, warmth, and security that is expected in any stable and safe village we might have had back on Earth. Only the setting has changed. We have more than everything we'll need to occupy our minds and our time. Deciding to leave our home planet is unprecedented, that's true. But we have the potential to make this new environment as tranquil and happy as any place down there and much better than some. It's all a question of attitude, of mindset. The scores of explorers who came before us accepted dangers just as great as ours, and many didn't survive the experience. All our testing, analysis, and knowledge points me to trust as fact that our equipment and planning are sound. We can ensure our physical safety more effectively than any covered wagon or sailing vessel ever could. I must again reemphasize that anyone can leave this ship at any time before we actually commit to commencing our expedition. But please be fully aware that I will also not permit any effort to abort the mission once we commit. The exception would be any situation where the ship and all its crew would be destroyed unless we did abort. All of that being said, we have yet to create our most important tool to progress toward the commencement of this mission. That indispensable tool is a formalized code of conduct by which we can govern ourselves going forward. This code will be enforced throughout our time on board and beyond, regardless of where the journey takes us. I propose that the code can only be altered by a two thirds majority of all crew members over the age of eighteen crewing the Conestoga at the time of the proposed changes. Are there any objections?" None were recorded. Adam continued.

"I believe the first question is how we should delegate authority amongst us."

"You mean like the command structure?" Peter Hosokawa asked.

"Yes, that seems a good way the phrase it," Adam said. "Mattie, you're recording this, correct?"

"Sure am," she replied. "That's my part of the command structure."

"Point taken. As I understand it, the plan we have in place now is to have a designated captain at the top of the chain of command. Currently, that would be me," Adam confirmed. "This has never really been formally stated, it's just how we've done things. Mattie has been first mate, again informally. Then, we have section chiefs, who are also currently informal. Manjeet has engineering, with Azaan, Gregor,

Iza, and actually Mary under him. Peter has the medical staff, Mila has astrophysics, with Jeter nominally under her. Then there's Sharon, and there's only one Sharon."

"Dully noted," Peter confirmed. This spelling is intentional. "So, the obvious question is, is this to be a permanent assignment? I mean, what if it turns out someone is actually unfit for their job? It may also turn out that someone is demonstrably not as competent in their positions in the long run as we initially expected. Also, what if someone is injured or killed? Does the captain personally decide how to replace them? Will the captain's word be the last word forever? I know that's how it works in the military, but it's not a perfect system in our case since this is not a military mission."

"Well, it is for Kevin over there," Adam quipped. "But if and when he comes along on this trip, he'll just be a military observer. He won't have any rank or authority over us unless we all agree to give it to him. Really, my thinking in this area is the same as I've been saying all along. We should all be cross trained as much as possible to take over anyone else's job on this crew if necessary. Mattie and I have been doing joint duty with all of you on the bridge for months now, and that's been going swimmingly, at least from my perspective. We want anyone in this room to be capable of serving as captain, or first mate, or whatever else needs doing. I therefore propose that all positions be filled by an anonymous vote, say, once a year, starting one year from now. I also propose that all assignments remain as they are currently, starting today if so voted. I suppose there should probably not be term limits as long as the people running are competent. Besides, assuming that everyone wishes to continue with their current responsibilities, there's no reason to tear them away from them at a time like this. But if people do want a switch down the road, this will allow that to happen. If someone is found to be lacking in motivation or job performance, then changes can be made."

"Between elections," Manjeet ventured, "it should still be the captain that makes the personnel changes if needed. Does that sound alright?"

"I think so," Peter said. "Unless, of course, it's the captain that needs replacing."

"I agree with both those statements completely," Adam said. "In that case, we would do what? Have a court martial? What do you think, Kevin?"

"I assume that protocols for court martials could be retrieved from our on-board database, Juni?" Juni nodded in the affirmative.

"I certainly hope nothing like that is ever needed!" she replied. "Actually, let me verify that for certain. Yes, it's all here," she said, looking at her laptop. "The uniform code of military justice is part of the main database. A prosecutor and a defense counsel must be appointed. Presumably, the rest of us would be the jurors."

"Would the verdict need to be unanimous?" Adam said, obviously interested in this information.

"I'd say a two third's majority would do to convict," Tariq suggested. "Justice is hard to come by where I come from, and I'm more than tired of that. On the other hand, issues like these have a way of getting factionalized. The yeas versus the nays can literally spiral into little wars. Getting a unanimous decision from everyone is nearly impossible. Two-thirds might keep everyone more concentrated on the actual issues at hand. Just an opinion."

"Works for me!" Mattie said. "Besides, I can't imagine we'll ever need it."

"We'll need to keep our bases covered for this Compact to work out right," Adam said. "Who knows what situations might come up over the course of our lives. We better be open and honest about these things while we have the chance. I move that we adopt Tariq's proposal for two-thirds assent for court-martial convictions."

Sharon seconded. The motion passed unanimously. Common sense Sharon took advantage of being recognized by the chair by calling on herself. "So, how do we actually press charges?" Sharon asked. "And not just for court-martials, for anything? And what laws will we base them on?"

"That's the real question, isn't it?" Peter agreed. Do we make up our own laws as we go, or use some legal code that already exists?"

"You guys are really digging into what I've been asking for, a serious consensus defining what issues we absolutely need to address. I thank you all for that," Adam said. "And as long as I'm speaking, I believe the compendium of United States statutes, or whatever the heck it's called, would work just fine. I believe the U.S. is pretty liberal as opposed to other nations, although I may be biased. Plus, it's what we've been using for the past few years back at Uplift. Juni, can you confirm we have access to that?" Juni confirmed that they could. "Ok then, any motions that we use that manual as our guiding set of laws?" There was a motion, a second, and a unanimous affirmative vote. "That's great," Adam said. I feel sure we won't have much need to use it. No one is getting paid up here, so there's no way to embezzle or swindle anyone out of their fortunes. No taxes to pay, either. Just don't kill anybody!"

"You never know what people might get up to," Sandra said. "I just hope no one ever steals anything from my pod, or hoards supplies, or gets into fistfights. Heaven forbid there's ever any assault, or rape, or worse."

"I definitely second that motion," Adam said. "Hey, what about laws regulating marriages? I knew I was forgetting something."

"You already said the captain could marry people," Mila said.

"True," Adam said. "Or whoever the captain is at the time. But do we need licenses or anything? Well, I guess we don't. But how about divorces? Are those allowed? Does someone need to formally give permission? Are same sex marriages allowed? For that matter, are they limited to only two people? I've said that this society of ours might need to stretch the norms a bit. If only for survival's sake, if nothing else."

"I think that some of us might be offended by flagrantly defying our social norms," Jeter said. "Quite a few of us, actually. Just be aware."

"But we've also said that one group of crewmembers can't realistically expect everyone else to agree, haven't we?" Peter said. "Religion, for example. One thing that would ultimately tear us all apart is that one religion was expected to be practiced by everyone. This is a multicultural crew, and that is by design. This time we definitely will try to go in peace for all mankind. All of mankind. And not just for appearances, for real. That should be written in our compact, Adam. No religious favoritism," Adam nodded in agreement. "Now, as far as marriages go, it should be just the same. You can be responsible for your own marriage but no one else's. Nothing can destroy human tranquility more than love and passion. I think if the married parties decide they're divorced, then they're divorced. No permission should be needed. The alternative would be locking people who can't stand each other into the same pod every night, living in hell. The parties would still pay a price, though. The price would be for them to stay part of the crew but not married. They can go back to separate pods and go on with their lives. They just damned well better get along with each other and everyone else when they do. Whatever broke up the marriage must be put behind them. Arguing and recriminating for the rest of the trip can't be tolerated. End of story."

"Yes, that's exactly the sort of problem I fear most of all," Adam reinforced. "It might well be some insidious contempt like that that could ruin us. Upsetting crew harmony would be something I personally would consider to be a crime if I were captain at the time. If you marry, marry carefully. That's all I can say. I believe I've done so myself.

"Yeah, you're good," Mattie said. "I should say I agree in regard to tolerance for cultural differences. Intolerance is the curse of humanity. If we wind up spreading it to other worlds, we don't deserve to be making this trip. We do what we need to do. If three of you men want to get married tomorrow, congratulations! Seriously! We'll have that much more love and happiness aboard. Of course, I know that won't happen, but those should be our principles from now on. No religious pomposity!" No record was made of there being any active dissent in regard to these issues. But there is no way for us to know if any adverse feelings to these statements were maintained among the more traditional members of the crew. At any rate, a motion was made and passed that marriages and divorces were to be decided by the concerned parties only and that sustained arguments between crew members could be subject to disciplinary actions by the captain. Disciplinary actions could then be appealed, if desired, to a court of the entire crew. A majority of that court could overturn the decision.

Adam voiced his own personal approval. "We've really done well so far with this compact of ours!" he declared." So, we embrace the general ethical laws of humanity as our guiding ideal, with tolerance as the prime ethic. Violations are to be judged first by the captain, with a trial by jury being an option if the decision is appealed. Charges can be laid by the captain or by any member of the crew.”

“I think maybe two unrelated parties might be better for laying charges apart from the captain,” Tariq moved. “Same reason as before. It might reduce abuse of the right.” Again, this motion was recorded, seconded, and approved.

“So, the next logical question is, what sorts of punishments would be acceptable?” Adam said. “Anything corporal seems pretty inappropriate. I hate the idea of maintaining a brig on this ship. We’re pretty isolated as it is.”

“I really have a hard time imagining anyone here would be capable of doing something so bad that they’d need to be incarcerated,” Mattie said. “I suppose all we could do would be just watch them closely. We’re all kind of on home confinement as it is. I guess we could do, what, confine them to their own pod if they’re single, or make a new pod for them by themselves if they’re not?”

“Maybe keep them in the Arboretum for the length of the sentence?” Xao Jia suggested.

“No, then they’d be in a general crew area for an extended time. Besides, there are also resources there that we must all have access to. A disgruntled prisoner might wreak havoc on the crucial resources we all need. If they were so inclined, that is. Best use a pod,” said Peter.

Gregor pointed out that prisoners needed guards, at least in theory. That would use up manpower around the clock that wouldn’t really be available or be fair to the guard.

“Oh, I could easily build a robot guard that could do that, no problem,” Iza said without hesitation.

"Well, let's write these options down, just so we don't forget," Adam directed. "I think we're more likely to forget what we've just decided long before we ever need to remember it. I hope if there's ever a need to implement any punishments, a generation or two will have gone by on this mission. All of us would need to discuss it anyway as a jury if one or more of us goes bad." Giggling was recorded with this comment. "I feel like I'm forgetting something, something important. What else should go into our compact?"

"Simple," Kevin said. "How about your conduct at the end of the trip, when and if we arrive somewhere worthy of our interest? Personally, I don't think I'd want to go anywhere if we screw the place up when we arrive, you know?"

"Yeah, that's what I was forgetting," Adam admitted. "Not a very good thing to omit. I don't even know where to begin."

"I've given this a lot of thought myself," Peter Hosokawa spoke up. "Can I take over the conversation?" Adam conceded the chair. "We can start with our own microbes, I think." A few sighs were heard, no doubt, for various reasons. "For starters, we need to look as closely as we can with every method of observation we possess to see if there is any life on any planet we orbit. If there is any, I say stay away completely and go home as soon as we can. Why, you ask? Did anyone ever see the original War of the Worlds? Remember what got the Martians in the end? Disease! They had no immunity to human diseases! At least, I think it was a human disease. It could have been bird flu, for all I know. But Kevin's right, we can't travel fourteen light years just to decimate any life we find when we get there. Thankfully, we probably won't be bringing too many viruses with us, we're all vaccinated against our usual human diseases and have mostly been quarantined for nearly a year without anyone getting sick. But who knows if some new virus is brewing up here that we can't predict? Then, of course, all of us here are already colonized with bacteria that we can't do without. Our intestines alone can't really function without bacteria. They help break down what we eat and make most of our daily vitamin requirements as well."

"Don't forget the soil we've brought aboard!" Sharon volunteered. "Sterile soil wouldn't allow our plants to grow properly either! There's a great deal of bacteria in our good soil. We'd definitely starve if they weren't there. We're all just immune to those kinds of germs, that's all."

"Correct," Peter continued. "But what if any life forms we might run into at Tau Ceti aren't? We could devastate their whole planet just by showing up!"

"Swell," said Victoria. "That hardly ever gets mentioned in the movies. It would ruin all the plots."

"Well, if nothing else, a whole lot of our time will need to be devoted to finding out what might be there when we first arrive. If there is anything, then we have to spend a lot of time determining if we can live with it and it with us. I think an ideal situation is to find a planet or moon that we can live on that has absolutely no life already there."

"But then it would have no organic matter in the soil for our plants and nothing in the water for our wildlife to eat. Maybe no oxygen either. "

"In that case, we could at least bring down what we have and live in enclosed structures," Adam said. If there's water there, we can break it down to oxygen and hydrogen like we do now. We can bring the aquariums down too, as well as the gardens. Maybe we could leave the seeds of life on the place, then go home and let someone else come back later. A little algae in the water can make a world of difference to terraform the place. We'd have to study everything in minute detail and form a plan then."

"And if there is life?" Sandra asked.

"We'd need to be careful and be sure that nothing we do there would adversely affect that life."

"How? We're humans!" Mattie said. "When has humanity ever not screwed things up? Wherever humans went on Earth, we always did, that's for sure."

"I'm afraid I have to agree with you there, Mattie!" Peter said. "We humans always seem to cause more harm than good. If there's intelligent life there, we'd need to be at least as stealthy as any UFOs that may have visited Earth."

"But our Shrikes are way less advanced as those probably nonexistent bogeys, "Tariq observed. If there's a culture there as advanced as ours, we'll mess with their heads in a major way. If they're primitive, we'd inadvertently make Gods of ourselves, and I truly hope none of us want that. We'll have to sick some of Iza's robocops on anyone who proclaims themselves to be a deity!"

"Ok, let's keep it real here, everyone. Yes, it seems that once we arrive, if ever, there will be far more reasons to not stay than there would be to remain," Adam asserted. "And if we need to leave, we certainly should as soon as possible, assuming the equipment is in shape to do so. Under no circumstances should we become unwitting conquerors or corrupters of anyone. Let's put that in our Compact. But on the other hand, if there is anything we find that we might consider to be wildlife like we have on Earth, I guess making it a food source would be OK. Let's just be sure we don't eradicate any species we come across. No extinctions due to our presence, please. Damn! Ethics can be so complicated!"

"And so vitally necessary!" Mila shouted. "All of these issues must be addressed. We mustn't forget that when the time comes! If humanity leaving Earth turns out to be the worst idea of all time, then we'll be the first to find out. It wasn't an issue on the Moon or Mars because there wasn't anything there to destroy except rocks and dust. We'll probably screw something up there anyway. But the point is, anything alive we find out there has as much right to exist as we do. We coexist, or we turn around! That's my feeling. Just discovering what's there is its own reward, not conquering it. Those days are past."

"Any dissent on this issue?" Adam asked. He was pleased to hear that there was none. It made everything much easier. "We don't have to discuss this further unless we actually do find something there that we shouldn't destroy. If we find it, then choose to leave, we'll have to worry that when they find out about it back home, they won't be able to refrain from coming back and screwing things up later."

"Whew! That's some major foresight you've got there!" Kevin said. "You know, I've really been skeptical about all this stuff you've been cooking up here. I know you've been kind of worried about me as well. I still am not one hundred percent sure about coming along, but I will swear I at least won't try to hold you back. You all appear to have the proper motives. Besides, I seem to actually kind of like it on this ol' boat."

"Glad to hear it!" Adam said. "I think you can see you won't ever be bored if you stay with us. I swear I'll tell everyone if a departure plan becomes imminent. You'll have one last chance to get off, Kevin, but that's it. Miss your chance and you're stuck with us. And for all our sakes, don't rat us out to your people. All of the rest of us want to go, if I'm not mistaken, and you have absolutely no right to tell us we can't. We came up with this technology, and we'd be going somewhere that no nation on Earth has a claim to. We can go anywhere we want if we're not even on the planet."

"Point taken. When will I need to make my decision?" Kevin asked.

"Within the next couple of weeks, I believe." No one gasped at this news this time. "So, Mattie, I think we have enough data to compose our compact. The Mayflower one was really just a few sentences, as I understand it. This one's a bit longer. Do you mind typing it up?" Mattie agreed. "We can all sign it before we go, if we go, and transmit it down to Uplift as a goodbye gesture. Holy Guacamole! We may just pull this shit off after all. Oops, sorry for the exuberance! This is what I was really waiting for, to hear everyone say we could pull it off together and that it seems doable. Is it advisable? Not really, but at least it's actually doable. Maybe we'll go into the history books together, successful or not. Either way, I couldn't ask for better company. OK! We'll meet again next week or sooner if I call you."

Mattie had innumerable talents, and apparently, crazy secretarial skills were one of them. She used her time on the midnight watch to refine and finish a final draft of this "Conestoga Compact." She then printed a pristine copy to be signed at the next meeting. Gregor and Manjeet confirmed two days later that the new mounts for the conventional rocket engines were installed and stress tested for the proper tolerances. They had decided that one engine would be mounted on each side of the main linear frame. When fired, which would be done with radio control, they would both be fired simultaneously, and half of the fuel supply of each would be expended. This would leave the remaining halves for any possible return journey if the mission was aborted in its early stage. Though conventional in design, the rockets used liquid hydrogen and liquid oxygen as fuel. These components weighed slightly less than earlier types of rocket fuel, making them somewhat easier to transport into orbit. But easy or not, the rockets finally made it to the station on the Wednesday after the compact deliberations were completed. Sparky had no problems facilitating the docking, mounting, and programming of the new behemoth pair of motors. Adam knew that another meeting was now needed with the engineers. This was called for the next day, Thursday.

He called out what was basically a checklist of go-no-go items that would be required before the ship could proceed with leaving Earth Orbit. Although checklists were a time-honored tradition in the world of aviation, they seemed a bit too sterile to Adam to be used on such a monumental occasion. The topics were listed as follows as per usual spaceflight protocols: "Hydrogen, oxygen, and nitrogen levels filled to maximum recommended requirements? "Go," Solar electric panels functioning at recommended levels? "Go." Nuclear reactor shells ready to receive fuel? "Go." All environmental systems functioning at recommended levels? "Go." "Argon gas tanks fully filed and without observable leaks? "Go." Antiproton tank containment fields functioning at specifications? "Go." All other fuel and coolant levels at maximum capacity? "Go." Any other concerns or systems not functioning at required levels?" Nothing but no's.

Juni confirmed the arrival of those twelve atomic battery-powered communication satellites which would hopefully help transmit data back and forth between the ship and the Earth using laser beams. A catapult would propel them backward behind the ship from a position on the outside of the big ring. This would hopefully fling them clear of the plasma contrails emanating from the center line of the ship but close enough for the small new sails recently added to catch to the peripheral contrails. Small thrusters, well, relatively small, would eventually slow the satellites down to a fixed line of travel where the data would find them and be magnified and sent backward down the line. Whether or not it would stop them at a fixed position was a big question, but this was not considered essential to maintain at least the minimal required lifeline that was desired. It also wasn't necessary that the equipment for this communication be completely ready before they left. The bots could mess with the relay satellites outside any time after leaving. They could even conceivably make their own additional satellites with the 3D printers. That piece of the puzzle was also now a go. No testing of the new missile system was thought to be required either. Though the platform and control systems were now installed, the missiles wouldn't be delivered until Mila went down to retrieve them. Even when they were finally installed, Earth orbit was not the place to do such testing. In fact, no one had thought of any way they could be tested as yet anyway. The team discussed the possibility of sending out some sort of target drone in the future and then wasting a missile to try to hit it. That would be have to be done sometime in the distant future after the ship had left orbit. At least it still seemed distant right now. Finally, Juni helped conclude the meeting by confirming that all readiness criteria were verified as fulfilled by the computer's A.I. This did not, of course, contain the few missing segments, such as the signing of the Compact and Mila's all-important trip to Earth. By this time, the four dedicated Shrikes that were to stay with the station on a permanent basis were securely docked at their respective positions around the big wheel. They were fully fueled and powered down, with the exception of their internal heaters. With a deep sigh, Adam gave the orders he had been hearing in his head for weeks now. "Mila, alert your people. I'm having one of the nuclear storage-capable Shrikes launched up here tomorrow to pick you up. The pilot is one of our best, Ana Kozelka, you know her. She doesn't know anything about our plans though, so watch what you say around her.

When we get the go-ahead, she'll take you straight down to complete your mission.

Chapter 16
A Rather Hectic Week

From meeting recordings and Uplift corporate records…

The third Monday of May 2088 was the beginning of a week of major consequence for the crew of the Conestoga. Mattie had finished writing the Compact, and everyone was strongly urged to sign. Signing was not an actual requirement for anyone wishing to continue on with the crew, but anyone who declined would be less likely to have their complaints heard and acted upon once the trip was underway. As it happened, no one declined to affix their signatures. This, of course, happened at the crew meeting begun at 10:30 a.m. that morning. Everyone had anticipated that it was to be an important one, and no one was late. Adam began the meeting by announcing that the Compact was now in final form, and written copies had been prepared for each individual or each couple in three of the cases. Mattie read it out loud and then passed the original around for signatures to be affixed at the end, where there was plenty of room to sign. Though the final draft will be included in the afterword of this chronicle, these are the names that were affixed: Adam Thorne, Matilda Orona, Manjeet Singh, Junaki Sato, Peter Hosokawa, Xiang Xao Jia, Tariq Al Abdullah, Sandra Annison, Victoria Zasco, Sharon Layton, Jeter Hamm, Iza Mazurkiewicz, Awamila Hayat, Gregor Levinski, Jeter Hamm, Mary Dupree, and Azaan Nazari. Kevin St. Patrick continued to decline to sign. The signatures were placed as the document went around the table and are, therefore, random. No order of rank or importance dictated where on the document they signed. Adam made a verbal note of this on the audio record of the meeting. Seventeen signatures. One abstention. Then Adam started speaking, but he was interrupted by a computer-generated announcement stating that

"Uplift Shrike number 182 is now docking at the main hanger bay. Pilot Ana Koselka commanding."

"I'll go greet her," Mattie said. "You have to open the access hatch from inside the station anyway."

"The co-pilot and Flight Engineer will be staying aboard," Adam said. They'll all be departing shortly.

"That's a bit unusual, isn't it?" Kevin said, sounding somber.

"I'll be explaining it all when Mattie brings Ana up here." Ana had only to let the Shrike's computer complete shutdown and then go through the pressure equalization protocol on the Shrike's airlock to allow it to be opened. She was wearing only the light pressure suit required at such times, which she was still wearing when the two entered the bridge. Ana had been aboard the station/ship numerous times during the ship's construction, but not at all since the construction had been completed. She was apparently impressed, and was beaming with a huge smile when she shuffled in. She had little time to adjust to the radial momentum and light gravity.

"Have a seat, Ana!" Adam grained. "It's wonderful to see you!" Since the usual conference room chairs were all filled, she set herself on one of the desk areas nearest the table on the bridge. "What did they tell you at pre-flight?"

"This is just a personnel transfer flight, right?"

"Yup, probably one of the last ones for a while, except when you bring Mila back up. Did you get any other details?"

"Not really. I know the flight computers got the trajectories uploaded, of course."

"OK, let me fill in the objectives, if you don't mind. They're a little unusual, I guess. Everyone else here needs to hear them too."

"I'm all ears," Ana said dutifully.

"Alright. You and your crew have come up to pick up Mila here and take her down to Uplift. I assume you've heard that we've sold four Shrikes to Pakistan. Mila is Pakistani, of course, and we need her to act as a translator for the mission. You'll be landing all five spacecraft, your own included, at Mascoor Air Base in Karachi. When you leave here, we'll be testing our reaction control motors for the first time and expect to enter a geosynchronous orbit over Pakistani airspace by the time you arrive there. We'll station keep above Karachi until you can return Mila back here to us by the end of the day tomorrow. As you may know, Mila is a bit of a celebrity these days in Pakistan. She already was before she joined us to do her work as a physicist. Now that she's an astronaut as well, she's even more famous."

"I'll be doing a press conference and giving a little lecture at Mascoor about our work here over the past year" Mila said. "The Minister for Air Defense will be there, as well as most of the defense wing personnel. You could watch it on Pakistani television here on the station if you can get the satellite signal."

"I'll work on that!" Juni said, only half seriously.

"Ok," After you're done there, the other four flight crews will get passenger flights back to Uplift courtesy of the Pakistani government. Ana, you get the lovely job of bringing Mila home. While you're down there, there'll be some cargo loaded aboard your ship that will also be delivered up to us. Once our robots unload it, you'll return to base. After that, our intention is to completely seal ourselves off from assistance from the ground for as long as we can stand it. It may be for years. Don't worry, though, we can still call you guys on your cells or maybe drop notes by parachute!"

"Yeah, we kind of have a pool going at Uplift regarding that question. I say you go two years tops, "Ana responded.

"Sounds like a sound bet," Adam said. "We'll just have to see, won't we? Any other questions? No? OK then. Pakistan awaits its hero!"

"Got it, boss," Ana said. "Good to see you, it's been a while. Say goodbye to all the fish for me!" With that, Ana and Mila departed for their Shrike, again followed by Mattie, who would see them off.

"You're really trying to pull one over on Nathaniel, aren't you? "Peter asked. "Or maybe it's us you're trying to fool."

"Nope, I'm done with that for good," Adam replied. "This is it, folks. This is the day. When that Shrike brings Mila back, we'll have everything we need to bug out of here. That will be the time to decide if you want to stay with this or just go back down to Truth or Consequences and forget you were ever here. Kind of an ironic name for a town, huh? There will certainly be consequences for whichever sort of truth you want to adopt today."

"What is Mila going to fetch?" Kevin asked tentatively.

"Well, some air defense missiles that are modified for operation in zero-G."

"Right, aaaand?"

"Two dozen nuclear batteries we can use for deep space communications satellites."

"Aaand?" Kevin said. "Goddamn it, Adam Thorne. She's getting us some nuclear fuel, isn't she? If you collected us all up here because you think we're stupid, you really screwed it up. It's truth or consequences time up here, too. A lot of us have had it with your equivocating excuses. I've been keeping quiet with my superiors up to now because I've believed in you this whole time. But this is a little much. You know damned well that private citizens can't just buy, what is it, uranium and walk away with no consequences."

"Yeah, you have been the only one of the crew that I've been worried about, Kevin," Adam said. "And believe it or not, I admire you for it. You've obviously got the ideal character for this kind of mission. You play everything straight. Yes, Mila's gone to fetch us enough plutonium to power two nuclear reactors for at least two hundred years. It's the only way to give us the juice we need to keep us electrically fat for at least a couple of lifetimes."

"But nobody has ever perfected a nuclear rocket engine ever in history. The risks are too great!"

"The same is true for antimatter drive. That seems kind of dangerous as well," Azaan said. This time, a true gasping sound was heard on the audio recording, no doubt coming from every non-engineer in the room.

"Oh, come on!" Kevin said. "That's the kind of thing that could really cause some devastation on the ground, not to mention us! Antimatter drive! Give me a break! How can you seriously tell me you've even come up with the antimatter? You guys are really world-class nutjobs!"

"Interstellar class, for that matter," Adam said. "You can believe it, everyone, we've done it. And we're ready to use it as soon as we can. But don't worry, we'll be more than a few thousand miles away from here before we light that candle. Earth won't feel a thing other than a lot of annoyance with UES Conestoga."

"UES?" Sharon asked.

"Yeah, United Earth Spacecraft," Adam said. "I made that up this morning. So, yeah Kevin, we'll be risking a lot, but so have all the explorers that ever lived on our world. Columbus' crew thought they'd fall off the Earth. Not to mention the sea serpents! Plenty of pioneers never made it to where they were going. Ever played Oregon Trail? What a bloodbath! Now, I'm not making light of this. Like I said, I'll go alone if I need to. But one more time. If it isn't us, it'll be someone else. Then they'll get the credit for everything we've done here all by ourselves and for the first time at that. I personally would rather die than let that happen. I know you think this is some kind of cult or something. I'm not sure about it myself actually. But at least I'm letting you decide for yourselves. This is it; this is the time when we get sucked into the heavens! If that turns out not to be what you want, then Ana can take you back down to Earth tomorrow. I implore you to leave if you're not sure if you want to come along. That way, if you change your mind after it's too late, you can't blame me. That's because there's no way I can really come back and get you after we leave. Understand that, because we all know it's true."

"You said that if the majority of the crew voted for something, that would override the captain's decision!" Peter said.

"Well, technically, we could slow down enough and return before reaching a planet," Gregor said. "But it would really be risky, take a long time, and a lot of fuel, too. I think it'd be just as well to keep going. Besides! We know how to live on this ship! It seems even easier than slogging around down there on the Earth every day. I want to go, don't all of you? We can die just as easily in a car crash or just flying around in our Shrikes day after day! We can make our own joy right here. No politics, less disease, free health care. The only thing holding us back is our own mindsets. Don't worry, Adam, you'll have at least one other person with you."

"When you all agreed to sign the Compact," Adam said, "I was under the impression that there would be more than two of us. Was I wrong?"

Just then, Mattie reentered the bridge compartment. "What? So, you told them about the antimatter engine?" she said. "I knew you should have done it sooner!"

"It seemed like a fantasy tale until a little while ago, even to me," Adam admitted. "Why get everyone all anxious about a dream?"

"And risk a security breach," Kevin said. "You're right, I would have told my superiors if I had heard. Funny, I admire people who can keep a secret. So, guess what? I do believe I would be doing better service for the Space Force if I tag along with you guys. It'll kind of be like T.E. Lawrence, remember him?"

"Lawrence of Arabia, of course, I remember him," Adam shouted. "You've all seen the movie, right? We should watch it if we ever have the chance. His general agrees to send him out to see how the Arabs are doing against the Turks. Well, He allows it more to get Lawrence out of his sight than anything else. He really hated the guy. So, while he's off on detached duty, he winds up winning the war in Arabia for the Allies."

"Detached duty, right!" Kevin said. "I won't ask for permission, but I'll inform the Space Force that I've accepted detached duty on my own initiative. I'll make it clear I wasn't coerced in any way, and I'll file reports on my activities if I'm able to do so. Anyway, they have no way to come up and arrest me, even if I'm charged with desertion."

"That's how I've been looking at it," Adam said. "And by the way, that's why we're about to move over Pakistani airspace."

"Why," Kevin asked.

"Well, technically, it's illegal to orbit nuclear material anywhere except over your own airspace. You can put a little bit on a satellite to power it, but it needs to be stationary over any country and only with permission. Flying plutonium overhead all over the world is the thing that's illegal these days. So, if Mila gets her cargo and meets us in stationary orbit over Pakistan, we're all good as long as we don't stray over any other country. When we're secure, and Ana departs, we'll light it up from right where we are sitting, over Karachi hopefully, using our maneuvering thrusters."

"Then we light those conventional boosters out there at the back, get some distance, and sayonara!" Peter said. "Merciful Buddha! This is what I've waited for and also what I've dreaded since I signed onto this project. But yes, Adam Thorne. I'm going with you." Everyone else agreed as well, including Kevin St. Patrick. Adam passed the Compact over to him, and the final signature was affixed.

"OK everyone, I'll be the only one communicating with the ground until further notice," Adam said. "Oh shit! I almost forgot, but it's OK. We still have until tomorrow to discuss this. For now, I'm about to inform Uplift that we'll be testing our maneuvering thrusters for the first time today. Those will be the only ones we need to change our orbit and move over Pakistan. There shouldn't be too much in the way of gravitational force change that we'll be feeling, but everyone should strap in somewhere when we do the burns. I'll announce it on the speakers about five minutes before. Get strapped down when you hear it. Now, tomorrow, it will be more intense. I suggest we all congregate in the Ed Givens for these burns. Gregor, could you install enough seats in her cargo bay facing forward? Get her cargo bay pressurized and secure the seats to the floor like we'd do for any launch. She's already facing forward at the top of the big ring. Ana will need the main docking bay to bring Mila back. I'll stay here to control the burn. I suppose I should halt the rotation of the wheels when we burn. That won't matter, right? Manjeet and Gregor nodded in agreement.

"Right, uh, captain!" Manjeet sounded awkward using this title when speaking to Adam." And once the burn is done, our acceleration won't be constant for a while, so we can resume rotation as our gravity source. We won't need to alter pod orientation until we have a constant acceleration."

"Yep, I knew that," Adam said. "OK, everyone, back to work. I'll be signaling Uplift about our imminent maneuver. Damn, I'd love to see the look on Nathaniel's face. Oh, first, I think I need the restroom! Good luck, everyone! I don't think you'll really need it yet, but Good Luck!

When the room had cleared, Adam moved to the communications station next to the computer controls. Moving in this part of the ship was always somewhat challenging. Even the outer portions of the small ring offered only about one-third of a G, and more towards the middle, where the bridge was located, there was always a pretty significant rotational force there as well. But once you were seated and focused on what was in front of you, it was something you could get used to. Adam generally spent a large portion of his time in this compartment and had come to think of it as like working in a swinging hammock or a floating raft. He eventually convinced himself that the sensation was rather pleasant; at least, that's what he would write in his log entries. The communication station was really just a laptop mounted on his desk. With it, he could access the Uplift control tower directly by visual and audio link. He spent a large amount of his time talking to the tower in the course of a normal workday. It was time to send a new message! This particular communication happened to be a bit different from his usual messages and also much more important.

"Uplift tower, this is UOF! (Uplift Orbital Facility).

"Go ahead, UOF." Adam couldn't put a face with the voice that answered him. He had been in orbit too long, he thought.

"Roger tower. This is Adam Thorne. Please be advised that we are preparing for a real- time test of the station reaction control system in t-minus 600 seconds. This will result in a decrease in spacecraft velocity and altitude with the objective of moving to an approved position over central Pakistan. This maneuver has been communicated to the Pakistani Air Command and will assist us in the recovery of crew member Awamila Hayat following her delivery mission tomorrow."

"Roger UOF!" the communications officer responded. "Mission Commander Kozelka has already filed a flight plan. Her Shrike in due to touch down at Uplift in 33 minutes. Good luck with your maneuver. We'll be monitoring it for you. Director Floatingfeather will be speaking to you momentarily. You seem to have piqued his interest. It will put you in the record books as the largest orbital object to change its position so drastically in Earth orbit, assuming you're successful."

"Thanks for your endorsement," Adam replied with syrupy sarcasm. "Communication completed." Next came the on-board announcement. "Conestoga, this is the captain," he announced somewhat awkwardly. "Orbital adjustment burn will commence in T-minus 520 seconds. Please find your way to any secure location nearest to you. If nothing else, hold onto something. A seat with restraints might be better. I'll start the verbal countdown at T-minus 60 seconds." He then made his way to the main bridge control screen, where Juni had preloaded the planned maneuver with the help of the navigation computer. A visual depiction of the Earth and the course line to be followed were currently on the screen, along with velocity and altitude data shown in numeric form. Adam had reason to be a bit nervous about this course change. The reaction control system utilized the same toxic gas that had killed his father, after all. Although it was all stored in tanks that were outside of the habitable spaces and had no chance at all of breaching the outer skin, it still gave him the willies just thinking about it. The jets had been tested, of course, but only for a few seconds to ensure they were functioning nominally. There was a copiously large supply of the Nitrogen Tetroxide for future use, and the actual amounts needed for precise small maneuvers expected on this mission were pretty minute. This would be the first time they were used, though. Therefore, Adam was still nervous. The computer showed all was proceeding as per expectations. At sixty seconds, Adam suppressed his jitters and went on the speakers to begin his count. It seemed somewhat overly dramatic to him, but it was not something he could skip.

"OK," he said, although that was not in the script he had in his mind. "Ten, nine, eight, seven, six, five, four, three, two, one. Ignition! We're firing!" Actually, these particular thrusters didn't actually "fire." They just expelled propellant gas to generate small amounts of thrust. No combustions were involved. But the effects were felt almost immediately. The station changed orientation very little, but someone had definitely put on some sort of brakes, making the station lose altitude and generally slow itself enough to lower its altitude. This made the Earth seem closer, which it actually was. Some lateral motion also moved the ship a bit more northerly. Although it could only be visualized on the computer screen, within about 20 minutes the orbiting spacecraft was seen to be approaching Pakistan. This was only because the computer image was kind enough to show national borders. Another round of rapid reaction control rocket firings and the ship was pronounced to be station-keeping.

"Maneuver complete!" Adam announced. "All crew report status. That's you're cellphones, people!" One by one, everyone said their names and "OK." Everyone was in fact alright. All reported no problems. Sharon added, "Were we supposed to be sitting down that whole time? Everything seemed fine after about five minutes, so I've been watering the rice. Was that a problem?"

"Apparently not," Adam announced. "It will be tomorrow, though. Don't forget!"

"Uh, Roger that, or whatever!" Sharon replied.

Just after that, Adam received a hail from Iza. It was regarding the four robots.

"I'm sorry I didn't bring this up before, Adam," she reported. "Just to remind you that all of our robots were plugged in and secured to their respective charging stations outside for our maneuver today."

"Yes, I assumed that was true, or you would have raised the issue during the meeting.

"Thanks for the confidence," Iza said. "Now that I'm thinking about it, I'm wondering how they'll do out there with a more substantial burn tomorrow. As you know, they anchor themselves in the charging stations using electromagnets in their feet, if you can call them feet. They can also grip surfaces with their feet, like primates. They can also do both."

"Yeah, I remember," Adam replied. "Don't you think that will be enough for the burn?"

"Well, probably," Iza said, sounding thoughtful. "But we really don't know what forces might be generated. I know it's not like we're in an environment with atmospheric resistance or anything. But we're going to need those bots a lot after we throttle up tomorrow. I think I'd rather have them secured inside, if you don't mind."

"Whatever you think, go ahead and do it. Sparky stays outside, I'm afraid. We can always make more bots, but why use the resources if we don't need to? You know poor Sparky's out of luck, though, right? He'll be carrying more radiation tomorrow than he ever has before. If he slips over the side, he's had it."

"Yeah, that was always a given," Iza agreed. "He's getting a bit ratty anyway, doesn't get to come in for a shave or a shower or anything. He's kind of gone feral out there. Best to leave him alone."

"Agreed!" Adam said. "Give him our best regards. He's going to be pretty busy tomorrow."

"Will do. Don't worry, the protocols are all ready to upload. It's not every day a robot gets to place fuel rods in nuclear reactors while in Earth orbit."

"Thank God," Adam said. "Take care, Iza. "We'll see you tomorrow."

Although everyone was understandably tense aboard the Conestoga for the rest of the day, no one was under more stress that day than Mila Hawat, who was not currently aboard. Her flight back to Uplift was short and uneventful. When she arrived, there was the usual crew debriefing in the hanger ready room. This was generally a boring and routine event. According to Mila, this briefing was a bit different, though. Nathaniel Floatingfeather was there in person to debrief her, as well as Ana and her crew. Nathaniel apparently really sensed that something unusual was going on aboard what everyone was still calling the UOF, and he grilled Mila about it in a way that seemed unusually aggressive. Mila played it calm. "So, Adam thinks he can just move his whole station right over the airbase you're planning to visit tomorrow? That seems a bit extreme, doesn't it? Any Shrike you launch from Pakistan can reach the station wherever it is. No need to move the whole thing there just to shorten the trip! I don't know what he's up to, but it sounds as fishy as one of his aquariums! What cargo are you supposed to be picking up? I hope he hasn't got all of you into a situation you'll all regret or that all of us will regret, for that matter! Come on, Mila, the last thing this world needs is a big dispute between America and Pakistan."

"OK," Mila said. "I'll admit we have made a deal with Pakistan to deliver some atomic batteries for the communication satellites we want to develop. We want to shoot them out into deep space to see if they can possibly increase signal speeds from locations farther out than Mars. It's just a proof-of-concept sort of thing. My country thought it might sweeten the deal to get us to sell them the Shrikes. Adam took it upon himself to agree. It's been almost impossible to get the US to allow us to buy anything with radioactive material, so he accepted an offer from Pakistan."

"And he thought it might be best to be over their airspace when he accepted the shipment? Makes sense, I guess."

"We'll be launching the satellites from there as well, so they won't enter any sort of orbit over the Earth," Mila asserted.

"Adam's gotten pretty savvy with skirting the rules, hasn't he?" Nathaniel said. "I hope he hasn't lost his grip on his self-control. In our business, it can get people killed pretty easily. The trouble is that these fly-boy types are always the first to crash and burn when they skirt the rules. Doesn't he ever worry you, Mila?"

"All the time," Mila said. "But someone has to do the visionary stuff every once in a while. Otherwise, the whole world withers. I don't like withering."

"Visionary, huh? Is that supposed to make me feel better? Somehow, it just doesn't, I'm afraid. Adam told me he wouldn't break any laws. This nonsense just squeaks by that promise, barely. Tell him I'll be watching him, Mila. If he crosses the line, he may be out of here. I can do that, you know. I've got the legal clout, and he knows it."

"Out of here, right! I'll be sure and tell him," Mila said with a face of stone. "Is that all, Nathaniel? I need to rest a little and finish preparing my presentation for tomorrow. Adam said there'd be a room at the dorm for me. Is that true?"

"Yes, Dr. Hawat," Nathaniel said with a nod and a minute little grin. "Here comes your driver now. Ana! You guys are off duty until tomorrow morning. You and the other crews get some sleep. And be on time for preflight!"

"Sure, boss, you got it," Ana promised.

And get it, Nathaniel did. At 06:30, May 22nd, the crews of five Shrikes, including Ana's, were at the ready room for their preflight briefing. Mila was on time as well, having been through this routine more times than she could remember. Actually, she had less orbit time aboard Shrikes than the majority of her crewmates. But nevertheless, flying aboard the big cruisers was one of her favorite duties. This flight would also be something of a novelty in that it would not be truly orbital. Though the little squadron would, in fact, skirt the boundary of what was defined as outer space, it would still be a direct flight to Karachi with only a brief period of weightlessness. No orbit would be achieved. This actually would be a drop in the weightless bucket for Mila, who most likely had spent more time in a weightless condition in various spots aboard the station than anyone else on the crew had ever accumulated by this point in their careers. Everyone donned their designated space suits, but there was no longer any need aboard the recently updated Shrikes to denitrogenate. Suits were worn using the usual atmospheric air mixture and were only needed in case the cockpit suddenly lost air pressure for some reason, which had never happened once in Uplift history. By 08:00, the five Shrikes were lined up on the tarmac, and the crews began their ingress procedures. The plan was for the ships to form up in a V as promptly as possible, with Ana's bird at the apex as she was the overall mission commander. Five spacecraft actually burning into suborbital space in formation wasn't something that was often done. It was a spectacular feat, but no one had ever found a way to film this maneuver effectively. Ana had a passing thought that Uplift ought to figure out how to do it. Not today, of course. By 08:30, they were given a go for their sequential launches, although technically, they were only really takeoffs.

Actually, such high-altitude and high-velocity flights were beginning to be seen as the future of long-distance air travel throughout the world. This type of methodology for intercontinental flight had been a goal of the airline industry for many decades, but the aircraft industry had paradoxically been slow to develop the necessary technology and infrastructure. Hydrogen as a fuel source, composite aircraft design, and replaceable heat shielding technology were all pioneered by Frances Thorne a few decades ago. Now these principles were well established and were just beginning to be utilized as the technologies of choice by the airline industry. This would be a slow process. Even the Aerospike engines that Frances made practicable were still poo-pooed by the majority of the aerospace industry. Actually, because of these prejudices and the Thorne's willingness to disregard them, Uplift seemed poised to dominate the orbital airline production market in the near and foreseeable future. It was an appealing business prospect. Mila would miss seeing all this come to pass, or so she hoped, as her small but fateful little formation assembled over Texas and engaged those aerospikes to shoot up to a high suborbital trajectory. The crews must have been a bit bored with such a seemingly mundane flight. The inter-craft communications were light and generally pretty jovial. A mere two and one-half hours were necessary to complete their transit. The airbase came into view rapidly, and the Pakistani Air Service welcomed them with a recently completed two-mile-long runway. Everyone on board the Shrikes hoped that it had been built to the recommended Uplift specifications for hardness and durability. The five birds landed uneventfully in sequence and were directed to taxi to a huge new hanger that had been decked out stylishly for the occasion. A large wooden platform had been built with the flags of both Pakistan and the USA well displayed. A speaker's podium was placed between rows of chairs, and numerous apparent VIPs were already seated facing a fairly large audience of uniformed Pakistani flyers. As promised, a few well-dressed people with cameras and microphones were also standing on and in front of the stage. The world press was here in force. The Shrikes were directed into the hanger and were powered down. Ana noticed that the small load of the promised replacement heat shields apparently had already arrived and were hanging vertically on the walls in special bays. She knew a big 3D printer was also scheduled to be delivered so that more of these could be made on-site as needed. More equipment for fuel production and storage would also be arriving soon. But today, it was the Shrikes themselves and Dr. Awamila Hayat who were the stars of the show. The last Shrike in line was directed to come to a halt outside the hanger at a moderate distance from the crowd. Although it had only been a suborbital flight, the bellies of these beasts were still hot enough to potentially cause some significant discomfort to those who stood too close. The crews disembarked and were seated in reserved spaces at the front of the seated spectator area.

The ceremonial transfer of the Shrikes to the Pakistani government was performed according to a script that Ana would read aloud, along with a video presentation on a screen behind her with dramatic footage of Shrikes in action. Ana spoke in English, with the Pakistani translation appearing in written form at the top of the screen. This was presumably done so that the American news media could broadcast it to the folks at home. The Pakistani Minister for Air Defense took the podium when the short film was finished and gave a speech that the English speakers didn't understand. Mila was on the stage, but otherwise, no translator had been provided specifically for the English-speaking visitors. The English subtitles were on the screen, but the screen was behind the English speakers and, therefore, could not be read unless the crews turned around. This seemed like that would be rude to the Uplift people, so they didn't really understand what Mila was saying. But from the enthusiasm of the Pakistani attendees, they generally seemed quite pleased about receiving the Shrikes. Ana Kozelka was presented with some sort of medal, or medallion at any rate. She'd bring it back to Uplift, no doubt, and put it in the cafeteria trophy case when she arrived. But first, the crews all had to shuffle to a small auditorium about a quarter mile from the hanger so that they could listen to Mila give about a two-hour lecture in Pakistani that showed enough footage and diagrams of the UOF that allowed the Uplifters to easily deduce what she was describing. This also became a splendid opportunity for the crews to catnap. Although none of them lived on the station, all of them had helped to build it in one way or another, even if it was from a cockpit or working a robotic arm. The physics related items of the presentation were not understandable to them regardless of the language they were delivered in. But the looks on the faces of the military personnel seemed to show they were in the same position.

Meanwhile, on board the secretly renamed Conestoga, the crew had been as busy as they had ever been since being awakened at 06:00 a.m. Mountain Standard Time. Adam had called an emergency but completely anticipated a crew meeting on the bridge. He informed the crew that Mila had called him and verified that she did not anticipate any changes in her day's itinerary and that the crew should proceed as though all of her goals for the day would be met. He told the crew that as far as he was concerned, the orbital exit burn would proceed as scheduled the following morning at about 10 a.m., May 23rd, 2088. The timing needed to correlate to when their particular point over the Earth's surface most closely faced their objective, and that happened to be the time. As long as they remained over Pakistani airspace, it should be alright to delay just a bit. His only concern was whether or not the Feds back in the USA had found out anything about the contraband coming aboard that evening. And if so, what might they be planning to do about it? He didn't include these fears in his morning announcement, but he probably didn't need to. He had no doubt that venting these fears would be the main topic of discussion among the crew that day. At the moment, however, there were hundreds of little jobs to be done before those big rockets were fired and they would be committed. Adam broadcast his instructions. "It is essential that all loose items in your personal spaces, as well as service corridors and accessible work areas, be secured as well as possible before Mila's Shrike arrives about twelve hours from now. Iza, please secure the bots, as we discussed. Medical staff, please secure your stations as well, especially the medication stores and fragile equipment. I know you probably have already begun doing so, but I need to record these instructions as an order in the ship's log. I'll be making rounds throughout the ship after breakfast, so let me know when you see me if you need anything. Thank you all. Stay calm and be vigilant. Captain Out." These last words seemed even stranger to him than ever.

But he also knew he'd just better get used to saying such things from now on.

There were indeed a multitude of worries to attend to that day. Sharon had been anxious for weeks about how her hundreds of plants would tolerate the jostling of a rocket firing. Would the plants be uprooted or get spilled out of their pots? She knew that at least the fish should be alright. Mattie had often told her about that first mission with Adam and the others. The fish would get plastered against the aquarium walls opposite the thrust but then recover as soon as the thrust was removed. It would be as if a small tsunami had passed over them, but their tiny fish brains would forget almost immediately that it ever happened. You couldn't actually ask them, but it seemed true as far as anyone could tell. The station rings' rotation would be halted only during the firing of the big rockets only, then resumed when the maneuver was completed. This would be repeated when the anti-matter drive was initiated, but that would not be attempted for a few more weeks at the very least. Only when everything was found to be as ready as possible would it be the time to rearrange the pods and other compartments to compensate for purely forward acceleration. Things should best be tied down now, though, to avoid flying objects hitting and damaging the station, or the crew. Gregor was busy early getting the seating installed on the Ed Givens. This is where the crew would gather when that ominous signal was given that would send them all into the vast void. It all was boggling everyone's mind at present, in spite of all the work and planning they had all been doing for the last couple of years. They all knew that this time it was for real, but how could it be, really? Their brains undoubtedly couldn't grasp it all, no matter how objective they tried to be. Everyone just did their best to shut out their thoughts and get to work. Was fear every sailor's curse? Probably. But there had been a multitude of sailors preparing to leave port over the centuries. The anchors had always been weighed, fear or no fear. So it would be with this crew. None of these sentiments, none of these dreads, were actually recorded in the logs or subsequent communications from the Conestoga. But everyone aboard must have had them on a day like this. One entry Adam did make was a communication with Mary Dupree. She was one of the quieter crew members and often didn't speak unless spoken to first. She always managed to do her job without much need for close supervision. He knew the various 3D printers on board were functioning in spectacular fashion. But before he decided to pull the greatest of all triggers on this greatest of all guns, he felt the need to check on Mary. She quickly answered her hail.

"Hey, boss," she said. "What can I do you for?" As always, she seemed quite low-key.

"Just felt the need to verify we have a healthy amount of raw printing materials aboard.

"Sure! What do you think I've been submitting all those invoices to Mattie for? Don't you look at those? Besides, it's way too late to do anything about it now if we didn't."

"Generally speaking, I trust Mattie implicitly. I guess I'm just as jittery as everyone else, you know? I'm trying to be sure I try to check in with everybody today, I guess."

"I appreciate it, boss. Is there anything in particular you don't trust me about today?"

"I was just thinking about those old wooden ships again. I do that sometimes. They always used to carry a few spare spares and rigging. Like ropes and sailcloth. I just wanted to be sure we had all our cargo secured."

"And you want to know if we have enough raw materials for repairs en route. Gotcha. Well, don't worry about that, Adam. We have small warehouses out there in the back full of borophene, aluminum, graphite, and everything that we made this boat out of in the first place. Sparky can whip up anything that's back there right now if we lose any infrastructure. Hell, he made it in the first place. He installed it, too. I don't worry for one minute about our manufacturing capabilities. And like I said, if we don't have everything we need, it's too late to get it now."

"Well, that's just what I wanted to hear," Adam said. "Now I only have a few other thousand things to worry about. Out of curiosity, what are you up to just now?"

"Helping Sharon secure all these friggin' plants, of course. It'll take most of the crew most of the day. Why, are you bored or something? You want to come to the arboretum and dispense some manure? Don't worry. It's human. Been processed and everything."

"I'm not scheduled for that until next week. That's one of my favorite jobs, though. See ya later, Mary."

"Don't I know it! Bye now."

This conversation was automatically recorded in the ship's log, as were all transmissions from the bridge. What Adam didn't know was that all such communications were being hacked by Uplift and made available to Nathaniel Floatingfeather. Adam had made sure that this was not the case for crew meetings or information entered directly into the ship's database. But anything broadcast over the ship's intercoms or crew cell phones risked being intercepted by Uplift. This had only been the case for the past two or so weeks and apparently hadn't revealed more than a whiff of a suggestion of Adam's overall intentions. This last communication with Iza wasn't a definitive statement that the departure of the Conestoga was imminent, but it was certainly suggestive. If anything, it showed that Adam Thorne was certainly a man filled with anxiety. The main thing was that Nathaniel still didn't seem like he had figured out Adam's intentions at this crucial time.

Adam was on the horn with everyone at some time or other during the rest of that day. He knew his crewmates well by then, of course, and they felt as though they knew him just as well. This level of interpersonal familiarity bodes well for the coming days. Nothing could endanger the mission, or the crew in general, more than distrust. People confined together in small spaces were always certain to find fault with one another. All you could do was hope that all of their faults were compatible with your faults. He knew everyone had mastered the hardware. Today he was making it a point to remind himself that this crew was truly ready to master themselves and proceed with the mission. He was also reminding himself that he had mastered himself as well. He decided that he and they were all actually ready, and that was the greatest milestone of all. The rest of the day was just as hectic and stressful as it was when it began. Sandra Annison, the G.P. made rounds throughout the day checking on how everyone was holding up. She recommended coffee breaks, and her personal logbook revealed that everyone on the crew received at least one dose of valium that day. Antianxiety agents were naturally included in the pharmacy manifest for a mission of this type, and Kevin had hoarded them in large quantities. Whatever medications that were dispensed were entered into everyone's computer chart. That data was still being transmitted back to Uplift, even at this late date.

While all this was going on in orbit, the time arrived when Mila and Ana Koselka's launch from Pakistan was set to take place. By Mountain Standard, and therefore also shipboard time, it was 1600 hours, 4 p.m. At the launch site in Karachi, the four transport crews had already boarded their jetliners for the flight back to Las Cruces, where it was currently 4 a.m. Flyers all through history had necessarily gotten used to oddball hours and thinking about how times varied all around the world. The Shrike crew, along with Mila, had turned in early the previous day. On the Conestoga, the crew had spent the last 24 hours catching what sleep they could, and this was short and sporadic at best. Down below, an abridged preflight briefing was all that was necessary, and the Uplift control center was notified that their launch was imminent. Mattie was on the bridge to confirm that the Uplift Orbiting Facility was prepared to receive them. Being just a short hop from the ground to the station nearly directly overhead, it was a mere thirty minutes from the 4:30 a.m./p.m. liftoff time in Karachi to the Shrike's rendezvous with the station. Ana soon announced her arrival at station keeping just outside of the main hanger bay. Mattie greeted her with a small update on the mission plan. "Roger Shrike 206," she said. "Be advised I am implementing a maneuver override to allow for off-loading of cargo at the stern robotic arm location. Please open your cargo bay doors. Our computer system will assume maneuvering control." Ana was expecting to offload her cargo but probably didn't expect it to be in the propulsion area. She no doubt assumed that construction way back aft had been completed. That was what was the official supposition at Uplift. Ana sounded a bit tentative on the com channel. "Can you repeat, UOF? The stern arm position?"

"Roger 206!" was the reply. "We feel this is the optimal offload position at this time. Computer control override in three, two, one seconds."

"Roger UOF, we're maneuvering now. Cargo doors opening." Ana sounded bewildered. The Shrike maneuvered into position within a ten-minute time span. Arriving Shrikes during the construction phase generally positioned themselves about 200 yards above the station just above the large wheel. Ana's only previous flights to the station had been during the construction phase. The top of the station was considered the side with the arboretum, as it was larger than the physics station on the opposite side. That was the structure they were maneuvering over now. The nose pointed towards the front of the station, where the bridge was. 206 backed towards the stern area and then lowered down to where the robot arm could access it. Ana looked out the window from the cockpit to the cargo bay. She had not inspected the cargo before taking off. This was something of a breach of Uplift protocols, but Mattie had specifically included instructions to forgo this step this morning. She sounded quite nonchalant when she said it, and Ana hadn't given it a second thought. Looking back at the cargo, though, she gave it seconds and thirds. "UOF! These cargo containers are marked as containing radioactive materials, please confirm. This was not mentioned on the manifest this morning."

"Roger 206!" Mattie responded. "I wasn't aware that you spoke Pakistani. That's pretty cool, Ana!" Ana was not known for her levity when in command of a flight, and she didn't break with tradition on this occasion.

This statement immediately resulted in another breach in protocol, this time from Mila. She interjected." You must not be paying attention to me when I reminisce about my home country! If you did, you'd know that there is no language called "Pakistani!" Our government uses Urdu, but that's just one of dozens of language variations and dialects. Those markings are Urdu!"

"Even so," Ana resumed, "You think I can't recognize international symbols for radioactive cargo, Mattie?!" she radioed. "This is highly irregular, to say the least. I require authorization to transport hazardous materials on all flights. Granted, this Shrike is rated for nuclear materials, but no authorization was received. I'll have to notify T or C before returning to base!"

"Roger that!" Mattie said. "You're not in any danger, remember. Your cockpit area is shielded, as you know. We will continue with cargo unloading procedures. Once this is completed, we request you proceed to the main docking bay to unload Dr. Hawat and refuel before disembarking. Actually, I've got the controls remotely, I'll proceed with these maneuvers now." Ana was audibly severely miffed by this major breach of protocol. She protested over the comm every chance she got from now until she returned to base at Uplift. When her taxi service Shrike docked, Mila escorted the Shrike crew to the bridge as quickly as possible in the spinning, low-G forward ring. Mattie and Adam were there to greet them. They were busy on comm with the rest of the crew. Since their stay was expected to be brief, no one but Mila took off their light space suits.

"Ana!", Adam beamed. "How are you? You look so, um, piloty! That is to say, you look very professional!"

"Just because I look professional, it doesn't mean I am!", she answered. "Look at you, you look professional, in charge even. But that appears to bear no relation to your behavior today. Not professional at all. Having me transport potentially illegal cargo aboard my ship and not even considering warning me in advance. That's not the Adam Thorne I helped certify as a Shrike skipper. I wish there was some way I could get away with placing you under arrest right this second!"

"I'm glad you're intelligent enough to know that that wish can't be granted here on my ship!"

"Ship!" Ana cried." This is a god-damned research station! And an American one, no matter what country it's flying over at the moment!"

"That's debatable!" Mattie said calmly. "Unfortunately, we don't have time for a debate just now."

"Why not, do you have somewhere else to be?"

"Sort of," Adam said. "Or at least, somewhere else we'd rather be."

"Too bad," Ana said. "If I were you, I'd sit down and listen to what I'm telling you."

"Well, I am me, I admit," Adam said. "But why don't you all strap yourselves into one of those chairs there, and I'll tell you where we'd rather be."

From the tone in Ana's voice, which began to change rapidly, she appeared to have had a small revelation regarding what Adam was starting to tell her.

"Oh, no…," she said. "I'm starting to think the rumors might be true."

"Really?" Adam said. "And who's spreading these rumors?"

"Nathaniel, of course, and anybody else with half a brain that works at Uplift. You're telling me that you want to take this so-called ship of yours out of orbit, right? That's why you're stealing whatever nuclear crap you have in those containers. Tell me I'm wrong!"

"You're certainly wrong about the idea that we're stealing anything. We paid for that cargo with the Shrikes we just delivered to those Urdu speakers down there. On their territory, it's completely legal. Oh, and look, we're still over their territory."

"Congratulations!" Ana pronounced. "But that's not the main question, now is it? Are you guys bugging out, or what? It'll be kind of obvious if you do!"

"Good point," Adam admitted. "Well, since you're all sitting down, I might just as well admit that, well yes, we actually are. As soon as we can after you leave, as a matter of fact."

"You know you can't do that. You always think everyone else is so stupid, you rich people," Ana shouted. "You know what I think? I think you and the Pakistanis are planning some sort of nuclear blackmail scheme. Who don't you like, the Indians, or is it the Chinese?"

"No, it's just like Adam said," Mattie countered. "And even if it isn't, there won't be anything you can do about it. You'll be forcibly escorted off this ship if you have to be. I know Mangeet could manage it pretty easily. And if you listen to that brilliant mind you claim to have, you'll realize that this is what we've been working on for the past two years or so. We just got lucky enough to figure out we could actually do it."

"To leave?" Ana said. "And where do you think you're going, anyway? Further than Mars? Why bring this big, bloated station anywhere else? Could you even orbit something this big around Titan or Io? Whatever it is, you should have gotten permission!"

"What, from Nathaniel?" Adam said. "He's the one that works for me! I don't need no stinkin' permission, not from him or anyone else." He said this in a surprisingly calm voice. "Look, I own this station, and I caused it to be built. We have a crew on board that's, well, on board with this plan. When you and your crew are sufficiently clear of this place, we're going to take her out and see what she can do. It's as simple as that."

Ana at last seemed to realize that this wasn't a joke. Her co-pilot and engineer, Debbie Harton and Micheal Wasson were their names, hadn't said anything so far. But they unexpectedly began to sympathize rather enthusiastically with Adam. Very enthusiastically, in fact.

"Why don't we go along with them?" Debbie said. "None of us have anyone at home that will miss us. How long will we be out?"

"It must be a few years, even, right?" Micheal asked.

"I'm afraid that we really don't know," Adam said. "It all depends on how things are operating. We might be back in a few weeks if we have to turn around right away."

"Fine, we'll arrest you then," Ana said. "Unless you land in Pakistan and ask for asylum. God, what a poor excuse for a boss you turned out to be."

"As I said, Ana, there's nothing we've done that could be called illegal under international law," Adam said calmly. "I know Americans think all the other nations of the world are subject to American law. Most Americans, anyway. But things don't work that way. I kind of wish they did too, but they don't. As long as we don't take our plutonium over any other nations, it's all perfectly legal the way things stand at this moment in time. Now, as for you all coming along, I'm afraid that's not possible. We are provisioned and trained to operate with the crew we have now. Not to mention that Nathaniel needs his Shrike back. Therefore, we need for you all to shove off ASAP. But first, I'm afraid I'll have to ask a favor from all three of you."

Ana was suddenly a bit more hostile again. "Oh great, now what? I suppose you think we'll go along with anything you suggest. Good luck with that."

"Since everything I've just told you is completely correct," said Adam, "I would like you to trust me when I say that after we've left, I will send a message to Uplift for Nathaniel to hear, for everyone working there to hear. I'll explain our motives and whatever plans we evolve after we've left. This means that I want you to promise not to try and stop us or impede us in any way. What would be the good of that in the first place? We're already risking our lives for the sake of exploration, like thousands of humans have done in the past. We won't be in any position to harm the Earth from where we are. You know we won't anyway. You certainly should." He opened the drawer in front of him at the captain's desk. "Ana, I have a letter here for Nathaniel. I'd like you to deliver it to him when you land."

"What's in it, if you don't mind me asking?" she asked simply.

"A sort of last will and testament regarding the company, Mattie and my stuff, that sort of thing."

"Will Nathaniel be the boss? The owner, I mean?"

"That's between us and him," Adam answered. "But you're certainly smart enough to make a good guess. There's one other thing, too. Something that involves you."

"Well, gee, and what would that be?" Ana asked.

"I think that if what we're doing doesn't end in absolute failure, and it might, then people at Uplift might get it into their heads to try and copy what we did. You know, why not try going to other places yourselves? Why stop now?"

"Could be, yeah. Sounds like my crew here might be up for it."

"My letter recommends that if there is another task force set up for that, then I think you ought to head it up."

"You're kidding, right?"

"Afraid not. Why do you think I handpicked you to be here at this particular moment? I wanted to tell you that I think you're the best qualified to keep all this going, except perhaps for your propensity for conforming to all the rules. You're kind of an authoritarian as well, but still the best choice."

"Is this to try to keep me from ratting on you all?"

"That too. But for now, just remember everything I've said. If everybody up here is willing to put their butts on the line, then I'm going to do it with them. If you want to stop us, then you never should have come to work at Uplift. You wouldn't succeed anyway if you tried. So, I believe your ship should be all fueled up and ready to go. Our flight window starts in about two hours. How about you all hit the road?"

Ana said little, just, "OK, boss. Good luck then, I guess, pilot to pilot." The crew was allowed to unbuckle and were escorted out by Mattie and Manjeet, who had just arrived. As their flight plan was already uploaded to the flight computer, they undocked, maneuvered away, and executed their deorbit burn within thirty minutes. Juni monitored their communications, and nothing seemed to indicate that there was anything exchanged with the ground other than the expected in-flight banter. Adam knew his crew was ready now. No other requirements remained for them to meet. He looked at Mattie, then went on ship-wide comm.

"Attention all hands!" he said, still feeling a bit silly saying it. "This is the Captain! This is an announcement that I have been waiting for a few years to make. Please complete any tasks you may be engaged in at the moment and make your way as quickly as possible to the Ed Givens and take your assigned positions. Mattie will be there to assist you. We will be in position to execute our departure burn 90 minutes from now, with a 60-minute period thereafter if needed to still be within the window. If we miss this, we could still execute our burn at this time tomorrow. This would not be optimal, however, as you all know. We'd be more susceptible to interference from the ground by then. Therefore, I will remain on the bridge to execute the burn as per our current operational plan. Please remember that aborting this burn will not be advisable once it is executed. So, everyone, please remain calm and proceed to the Givens as soon as possible. Captain Out!" And there it was. The die was cast, whatever that meant. Considering all the work, the damned, damned hard work that had gone into reaching this moment, it all seemed sort of disappointingly routine. They all were being asked to try the impossible once again, just like all those other times that got them here. And like all those other times, they got on with it without questions or complaints. Could they do it? They had asked all those questions before and knew the answers ahead of time. They were leaving now, just as scheduled. Of course they were, no doubts remained. They were leaving, but they were bringing their home with them when they left. For all of us left behind, it seems strange to think about it that way, but that's the way it was. For the crew, it was more like the Earth was leaving instead of them. They'd miss it, but it was the rest of humanity that decided not to come along with them. Too bad for Earth, but humanity would just have to deal with it. At least, that's the tone Adam would take in his logbook. This countdown would proceed, no questions asked.

Chapter 17
Where No One May Go Again

Mattie was already on her way to the Ed Givens when Adam made his portentous announcement. This particular Shrike was chosen because it was already oriented in a T position in relation to the big wheel, with the cockpit windows facing towards the front of the ship. Everything in front of the big wheel was smaller in diameter than the wheel itself, so there was an unobstructed view of what lay dead ahead. More importantly, the thrust from this burn would be forward, and the G-forces would be eyeballs-in. This was the traditional preferred position for any introduction of forward force. Everyone would "merely" be pushed back into their seats, not in some less comfortable orientation. Once the two conventional rockets lit (It was decided that they would be burned together to balance the thrust forces equally), the total forces would equal about four G's. At least, that was what was projected in simulations. Another bonus would be that both engines would still have half of their fuel loads leftover in case further burns were needed later. Nothing entirely like this had ever been done before, which added another layer of tension to the maneuver. The burn would last about four minutes, enough to start them on a trajectory that would probably need very little in the way of correction later on. Since they would not be traveling along the same plane that the solar system occupied, planetary gravity would have little in the way of effect on the ship that would corrupt their chosen path. As the rest of the crew floated aboard the Ed Givens, they tended to float towards the ceiling of the Shrike as it rotated around the outside of the big wheel. One by one, they entered through the airlock and then worked their way to the closed and pressurized cargo bay. Here, they each chose one of the seats for themselves, all of which were bolted to the floor below them. Flipping around and swimming against the grain of centrifugal force to get strapped in was more challenging than many wired nerves could take. Tariq was the first on comm to protest.

"Mattie, when you reach the apex of this rotation, could you have Adam shut it down? I think it'd be easier if we just had zero G back here from now on."

"Roger that, Tariq. You hear that, Adam?"

"Understood," Adam said. "It was going to do that automatically two revs from now, but you guys would know best. It'll stop now when you reach apex." Apex meant when the Givens was in line with the center of the Arboretum.

"Many thanks," Tariq acknowledged. Everybody on board was now wearing a headset and microphone. It was the same setup they would be using if they were all in pressure suits, which they were not. The crew had decided during their frequent chatting today that full-on flight suits would be superfluous. Shrike cargo bays never depressurized accidentally. As promised, within a few moments, the rotation ceased, as it did on the small wheel. Singly and in pairs, the entire crew presently floated in. Mattie, Mangeet, and Juni were strapped in as Pilot, Copilot, and flight engineer up in the cockpit. From this vantage point, the two pilots were able to monitor the Shrike's systems, but this was just to keep an eye on basic functions like power, atmosphere, pressure and temperature. Juni, however, could monitor the whole of the Conestoga from her chair, with computer access and control override capabilities on the odd chance that this data happened to be needed. Adam, of course, would control the burn procedure from the bridge. There was, in fact, a virtual pilot's seat in front of the room on a big flat screen facing forward that was configured to look just like the pilot's seat on any Shrike. Theoretically, if there were an actual window in front of him or whoever was piloting, they would now be able to see the star they were shooting for with their own eyes. Mattie could as well in the Ed Givens. Considering the number of other stars in their field of view, however, they might be hard-pressed to pick it out. It was enhanced and circled on their flatscreens though, and it reassuringly seemed to be about dead center of the screen. This was misleading, though, in that at the moment, the Conestoga was flying forward in its circular orbital path around the Earth. The screens would show the target to be reached more clearly once the RCS engines had pointed the big ship directly at it. There was still a large window of time before all this maneuvering could commence. Waiting was hard.

The whole crew was assembled and strapped in about 20 minutes prior to the scheduled optimal burn time. Adam, Mattie, and Juni began proceeding with their pre-burn checklists. The rockets they would fire were of the tried-and-true conventional kind, except that the fuel to be burned was liquid hydrogen and not more conventional rocket fuel. Even so, the checklist now being recited would have sounded familiar to anyone who had lived over a century plus two or three decades ago. It all sounded hauntingly similar to the old Apollo countdowns on television. There were verbal checklists regarding fuel pressures, indicator lights, and gyroscope settings. Even astronauts of the late 2080s still preferred to not let these important steps be executed automatically. They still demanded that countdowns be controlled and verified by human eyes and fingers. This was especially true if one was not actually sitting in a control room on the ground. It was not recorded what the crew members sitting on that Shrike were thinking and feeling while all this was happening before their eyes. It was true that anyone who had worked with Adam and Uplift for several years would be used to being on a spacecraft during a launch. But quite decidedly, this launch was different. It was undoubtedly the largest spacecraft ever to be used in such a way, for one thing. But the ominous nature of the flight to come was a great weight resting on their shoulders in the weightlessness of the cargo bay. It couldn't have been the case that anyone was in a lighthearted or jovial mood. But remarkably, there is no record of anyone panicking or suddenly yelling that the countdown should be stopped. Every person strapped in those seats had been told that if they wanted to leave, Ana Koselka was their last taxi home. No one took that taxi. Nevertheless, it seems hard even now to envy any of them, although many do, even now.

Though it felt like this long day should have lasted much longer, it really seemed like it had lasted no time at all. At t-minus 15 minutes Adam was heard hailing the control tower at Uplift. "Tower, this is UOF!"

"Roger, UOF, this is Uplift Control."

"Be advised that we will be attempting a maneuvering burn in fifteen, that's one-five minutes to accomplish a trajectory change."

The tower, of course, had no idea of the magnitude of this trajectory change. They appeared to not sound terribly surprised. Even so, they seemed to appreciate the heads-up. "Roger UOF!" was the only reply. Adam heard his crewmates giggling loudly on his headset. He wasn't really sure what they were laughing at. He most assuredly wasn't laughing himself. The countdown continued, with banter going on thick and fast between Adam, Mattie, and Juni. Adam turned on the camera in the Ed Givens cargo bay. He saw that everyone was calm, although many had their eyes closed and appeared to be praying or perhaps were simply rocking and talking to themselves. He asked whether everyone had had a chance to use the restroom before reporting. "Shut up, Adam!" was the nearly unanimous but not simultaneous reply. "Five minutes!" then four, then three, "Remember everyone, this first RCS burn is just for maneuvering the ship to the correct orientation. We're just gonna wheel around now to point ourselves at Tau Ceti."

Mattie and Juni announced that all was proceeding properly. By then, the countdown was nearly completed. "Five, four, three, two, one, RCS rockets firing! Adam said calmly. Twenty small maneuvering rockets fired at once. This was quite a large number in comparison to anything that had been required before on all those other, punier spacecraft that had come before. The sensation must have felt similar to a large ocean liner making a sudden emergency "hard left!" when some unsuspected hazard was sighted by the watch. It is not known how many of the crew had ever even been on an ocean liner, but the experience surely must have been different than anything they had experienced before. The maneuver that got them over Pakistan certainly should have been a bit gentler, changing their velocity and orbit, but this was certainly not the same as wheeling the whole ship to an entirely different orientation. Things must have really started feeling serious now. Mattie was heard to start singing softly to herself. "For those about to rock, we salute you!" the flight recorder chronicled for posterity. Mattie's voice had a lovely pitch that day. Once again, a go-no-go for launch was broadcast in Adam's much less lovely voice, this time for t-minus five minutes. Juni's navigation subroutines were taking sextant and other readings to verify that the ship was oriented properly for the burn to commence. Ship's systems and rocketry status were verified and re-verified. No halt in the countdown was recommended. Adam saw the expressions of his crew were as intense as he ever had, or hopefully, ever would see them again. Most appeared flushed, and some were sweating profusely. He made himself recite the lines.

"Orbital departure burn in ten, nine, eight, seven, six, five, four, three, two, one, ignition! We are burning!" He felt, as everyone did, that "cosmic kick in the pants" that could only be a pair of supremely powerful rockets being shoved up everyone's bottom at the same time. Spacecraft seats are certainly built to absorb a lot of this, but it is doubtful that anyone who has ever experienced a major burn has had nice things to say about it. Oh wait! No one in history had ever had an experience like this before. All we know is that no one appears to have been injured, as no one appears to have reported to sick bay afterward. The initial kick was the worst, but the G-stress was increasingly uncomfortable as the burn progressed. Azaan reported through the strained breathing that none of the engines he had designed and built would be so brutal, assuming they would actually work at all. Juni reported at regular intervals during the burn that all systems appeared operational. External cameras were showing that no damage, or even potential damage, could be observed. They had built themselves a tight ship, all right. Even Sparky could be visualized, and he seemed completely unimpressed by the ride.

The burn was scheduled to continue for approximately four minutes. This was calculated to be adequate to cause the Conestoga to escape Earth's gravity so effectively that it would take the ship over a month to slow down and be recaptured again to enable the ship to turn around and make another burn to get back. The deployment of the solar sails, and also the activation of the ion drive might make this process take less time, but these things were not on anybody's mind just now. These were contingency measures, but for now, the engineers were entirely confident that the anti-matter drive would check out well and could be activated very soon after leaving Earth orbit. To evaluate what had happened to the ship by the end of this burn seemed challenging in its own right. The Conestoga was quite large in comparison to anything else tossed away from the planet before today. At least the Conestoga was being operated in a vacuum, and there was no significant air resistance for the ship to push against.

While conventional spacecraft modeled after big rockets like the Apollo Saturn Fives, had only used their biggest engines for the purpose of hoisting the comparatively minute payloads off the ground. The Conestoga was purposefully made with the lightest materials possible but nonetheless was filled with huge amounts of liquid water and liquified gases in addition to her vast amounts of equipment. The human beings were trivial in terms of mass, but there were also four Shrike Spacecraft attached to her as well. There were no scales in space, but a low estimate stated that her weight, mass actually, would be equivalent to over one hundred fully loaded Saturn V's before they were even launched. Now the Conestoga had the equivalent thrust of two Saturn first stages pushing her away from Earth orbit, not from the Earth's surface. When leaving Earth orbit, the little Apollo service module engine's max estimated cislunar velocity was about 40,000 kilometers per hour and could fly from the moon to Earth in a bit less than three days, about one-quarter of a million miles. Imagine how far away this Conestoga could travel in one month once Azaan's new antimatter engine was activated. Even so, it would initially be nowhere near the half-the-speed-of-light kind of velocity that would be necessary to get this crew where it wanted to go within the hoped for twenty years. But they had to start somewhere before they could even get the thing far enough away from Earth to safely try it all out.

Even before this conventional burn was finished, Adam was being hailed from Uplift, near the city of Truth or Consequences, New Mexico. "UOF, UOF, come in, please! This is Uplift Tower! Our scanning indicates that you have left orbit at a tangential vector from Earth and at great speed. What is your situation? Over. Do you wish to declare an emergency? Over."

Adam had been prepared for this sort of response from below. He was hoping he was just being overly pessimistic. But, yes, he had been afraid since the beginning that some ominous party on the ground, most likely the United States government, would actively try to stop his unauthorized departure. These fears led him to continue with his planned ruse. "Uplift, UOF!" he announced. "We have ignited our two rocket engines in order to make measurements of our thrust-to-mass ratios as per our research goals. Apparently, there was an error in our computer's vectoring subroutines. We are not on the trajectory we had initially planned."

"Understood UOF!" came the reply. "Repeat! Do you wish to declare an emergency?"

"Negative, Uplift," Adam responded calmly. "Our spacecraft has suffered no apparent damage or loss of integrity. We still have 50% fuel remaining in our main thrusters and can reverse our course and return to our initial position at our own discretion. The stress analysis can still be completed regardless of our course. We will contact you as soon as our action plan is developed. Do you read?"

"Um, Roger!" the woman on Comm responded. She was tentative in her speech, as though she were taking instructions from someone as she spoke. Sure enough, a more familiar voice replaced hers on the mike.

"Adam, this is Nathaniel," said the new voice. "Do you read?"

"That's affirm, Nathaniel."

"I've got Ana Koselka here with me, and she says that maybe you have more happening up there than a miscalculated trajectory. Is there anything else you should be telling me here?" There was a short delay before Adam responded.

"Ah, negative Nathaniel," Adam radioed. "It doesn't matter anyway, we're all squared away here. The ship and crew are in no danger, and we can make the planned evaluations as scheduled and report back to you by tomorrow, over."

"Very good," Nathaniel said cautiously in response. "We'll be waiting to hear from you. And by the way," he added. "Ana here tells me that in the event that if some change in your originally filed flight plan occurred today, that somehow you thought we'd try to shoot you down. She says you thought we might even try to arrest you or something. Now surely you must have figured out that Uplift doesn't currently even have the capability of doing something even remotely like that. I admit that this is rather disappointing from a personal point of view, but it is a fact. Please don't lose any sleep over it. You'll probably need all the sleep you can get tonight as it is. Now, as far as anyone else taking potshots, I think you must be nuts to think anyone cares about you that much. Besides, they've never heard of Ana and probably believe that what you just broadcasted is true. We haven't even heard a peep about your leaving in the foreign press. Most of the world is asleep right now, anyway. So, you all just take care up there, OK? Let me hear from you when you've got your story figured out. You're a strange bird, Adam Thorne. Lucky for you, I like strange birds. Uplift out."

"Roger that," Adam said softly. He switched off his comm, then went on the ship's speakers. "Attention all hands! My screen confirms that we have had a successful burn and that we have achieved our expected trajectory and velocity. Our thanks to the Uplift tracking system, which we are currently tapping into. Engines are shut down and stable. You are all free to exit the Ed Givens and to return to your stations. Please examine your duty stations for any damage or other problems. Once this is completed, please sign off with Mattie and then return to your quarters to start your sleep periods. Meet me tomorrow at 9 a.m. for a crew status review here on the bridge. This is a bit curt, but congratulations to all of you for a successful burn. Captain out!" And that was all there was to it that day, May 23, 2088. If there was something more consequential he could have said, he probably would have, but he couldn't think of anything just then. A small group of humans had just departed from Earth with the intention of trying to reach a star that wasn't theirs, but no better words came to him. Everyone was probably just glad that they hadn't blown themselves into oblivion, etc., and that would do until the next time they attempted their next maneuver. Anyway, it had been a very, very long day, and adrenaline could only carry you so far. They were past that point already, but duty called, and they spent another two hours checking the ship. Then they all went to bed. Adam slept on the bridge, as that was a Captain's duty. He announced that he was reactivating rotation for both habitation rings, then floated off to his hammock/bed.

His alarm went off at six am as usual on May 24th, 2088. Adam realized that this meant he had only gotten about four hours of sleep that previous and nearly incomprehensible night. Mattie came in just after the alarm. She looked tired, too, but was also strangely bright and bubbly. Good morning, mon Capitan!" She giggled. "Sleep, OK? I didn't really, but I don't appear to give a crap. Can you believe what we just did? Did you have Nathaniel on visual last night? What did his face look like? Did you record it?"

"No, no visual," Adam yawned. "But the audio, yes. I always record everything that happens on this bridge, you know that. Hell, we're being recorded right now."

"Yeah, I know," she sulked. "When are we going to really celebrate, then? You know, as husband and wife?"

"When everyone gets here I'll have to ask that someone take the next night bridge watch."

"Or command someone to do it."

"That sounds right to me. Good plan. You know what? How about I text everyone to hold off on reporting until 10 a.m.?"

"Wow! A whole hour later! You're just a saint. Are you saying we should celebrate early? You can't leave the bridge unmanned right now, you know."

"Yeah, I know it, and I damned well can't turn off the log recorder either. I'm only kidding. We'll just have to celebrate some other time. Any system alerts last night? He asked.

"Let's see. Well, there was a 15% drop in battery charging from the solar arrays since our burn. I'd say we spent a bit more time in the Earth's shadow than usual, don't you think?"

"We can ask Mila, I guess, but things were a bit screwy after our first maneuver. The ship's never been moved around this much before. Probably won't happen again for a few decades."

"Let's hope!" Mattie said. "So, what's on your mind for today?"

"I've got a mind to do a conference call with Nathaniel. Maybe we should just level with him about what we have in mind."

"You know, I personally think we should get a better idea of our status first. Maybe we should get the solar sails up and be sure we don't have a leaky boat before we commit. We'll look kind of dopey if we say we're leaving the solar system and then have to turn around in a week or so. You gave Ana that packet, right?"

"Yeah, she took it. Actually, all it contained was an access code to my safe at home. I faxed a copy of our will, some requests, like having Ana take over the new task force, and some financial information down to Frieda Jojola. Actually, I'd like Frieda to take over the Thorne Foundation for the charity work I've endorsed. I also included the letter Kevin wrote and signed. It's a letter of intent to the Space Force saying he was not resigning but that he considers himself on detached duty."

"Nice!" Mattie agreed. "Now, as I recall, you have to signal the house with a second access code before the safe opens. Yeah, you do, I remember. I don't know how far we can be before the signal stops being timely. Maybe you'd better just do it today."

"That's what I'm thinking. You know, why don't I text the other code to Frieda today and have her bring Nathaniel and whoever else he wants there to the house. Then he can be right there when he gets the deed. A little complicated, but I don't think we'll need to wait too long."

"Hombre, leaving the planet was complicated! This other legal stuff will be a lot easier."

"Mrs. Thorne, you knew the job was dangerous when you took it!"

"Oh, quoting Super Chicken, eh? That's a strange reference to make at a time like this! Maybe I'll start calling you El Pollo Grande!"

"Probably more accurate than Captain, but don't tell anyone. So, that was the only alert?"

"Well, just the usual temperature fluctuations throughout the ship, but the heaters were able to compensate. Here, why don't we just bounce over to your hammock over there and equalize our own temperatures?"

"One giant leap for wife kind! OK, c'mon!" Mattie texted the delay in the meeting start time. One can assume the Thornes took an extra hour's nap as well.

At about 09:45, Mattie assumed the bridge duties, and Adam made his way to the adjoining bathroom to change and wash up. He did manage to get enough sleep overnight to function adequately today. He wondered how the others slept as well. Sandra had dispensed plenty of sleeping medication even before the crew had gathered on the Ed Givens. By 10:00, the usual parade of crew members floating and bouncing in had finished. These people were punctual.

"So, how did your first night as expatriates go?" Adam asked.

"Strange to say," Iza answered, "but it really wasn't any different than any other night we've stayed on this ship. Our routine was upset a little, but we were still home." Everyone else indicated that they had felt pretty much the same as Iza. It was usual for every pod dweller to close their "curtains" to the outside at bedtime anyway. This was so the light on the day side of the Earth didn't alternate with the dark of the night side as they orbited. That's what usually made it hard to sleep. With the curtains, or more precisely a blind like those on airplane windows down, there was no indication that they weren't still near the Earth anymore.

"So, what's on the agenda for day one?" Peter was the first to ask.

"Well, certainly proceed with your usual inspection duties," Mattie said. Assignments were still her department as first mate. "If ever there was a time for things to get knocked around, it was yesterday. We need to be sure nothing was damaged or rendered nonfunctional."

"I agree there!" Adam said. "And let Mattie know when you feel you've done a good comprehensive check. Report any damage, of course. But when that's all done, I think we need to get those solar sails deployed. I'd like to start getting that extra boost as soon as possible! That's for you engineers to do and the robots!"

"Oh yeah!" Azaan giggled. "A few extra kilometers per hour never hurts."

"We take what we can get," Manjeet agreed. "It all adds up, you know."

"I assume the protocols are all loaded on the bots?" Mattie asked. Popeye, Blackbeard, and Sinbad would be present to assist with unfurling the solar sails if anything went amiss with the procedure.

"What do you two think I do all day anyway?" Iza sneered. "My little guys out there need a lot of TLC, you know. But actually, I prefer to garden most of the day. My three musketeers outside are ready anytime, though. You know I made up some little rocket packs for them a couple of weeks ago, right?"

"Hmm, I don't remember you saying anything about it to me…Oh wait, yes, I do. Well, anyway, remind me again." Adam said.

"Well then, your wish is my command. Since we're not doing any accelerating or maneuvering currently, jet packs seem about as reliable a way for the bots to get around out there with some little peroxide thrusters as it is for them to hook onto rails with their fingers and toes. Why walk when you can fly, right? Of course, they will still anchor on when they reach their designated workstations, but this should help things go faster."

"I never question your work, Iza," Adam said. "And I'm all in favor of pushing on full speed. Let's get those sails up as soon as we can, as long as it's safe."

"Aye, aye, Captain," Iza said in her usual semi-reverent way."

"When that's accomplished, then we will go to the ion drive?" Azaan asked.

"Well, yeah," Adam answered. "You know we engineers like to proceed one step at a time. Analyze the last step thoroughly before proceeding to the next one. We want to make sure the sails are having an effect. Remember, they're designed to collect any protons, antiprotons, or plain old Hydrogen they come across, too. Well, Azaan remembers, of course. But anyhow, we should get some idea if they're working before we, um, continue moving forward."

The crew was then dismissed, and the first full interstellar workday in human history commenced. Sharon had suggested that the crew assemble in the baking area that evening at six p.m. for the first feast ever to be held in this part of the solar system. She had taken a frozen turkey out of storage and let it thaw. She had actually done this once before for Thanksgiving, so it wasn't a completely novel idea. But the gesture was well-earned and well-appreciated. If there was ever a fitting occasion for a celebration, their first day on this cruise certainly was one. As the crew feasted, the robotic three out of four nonorganic crew members working outside the sealed biosphere toiled away. The huge areas of the two solar sails began their unfolding processes. This would be completed in a mere two days, counting the establishment of control and power collection functions. Sinbad signaled the completion of this event in his odd Arabic-sounding simulated voice over the ship's loudspeakers at about six pm on May 26th. By this time, the crew was thankful to discover that living aboard the Conestoga when it was not orbiting the Earth was not too terribly different from living on it when it was. The following day, Adam decided that he had better have his conference with Nathaniel sooner rather than later since the ship was, in fact, moving away from the planet at a pretty decent clip. To wait much longer would create long delays between transmissions and replies. That would inconveniently make everyone involved in the call more irritable than they undoubtedly already were. The computer directed that North America would be in the proper facing position about 9 p.m. the next day, so he sent a transmission to uplift the evening before to ask everyone to be ready. He asked the appropriate people, especially Nathaniel, to gather in Adam's private office at the mansion. The crew was also asked to try to be on the bridge at that time. This was where the main communication equipment was located. As an encouragement, he promised that this was likely to be the last of the full crew meetings he expected to call for the foreseeable future. After this, a more convenient and less formal spot would be selected, and the meetings would generally revert to once each month. For this one, everyone was able to gather at the specified time, and the mood seemed quite light-hearted. Adam dialed Nathaniel's phone at 9 p.m., and the call connected about three minutes later.

"Adam! Is that you?" Nathaniel answered.

"Sure is," Adam replied. Wouldn't that be annoying if someone else decided to call you now? It would suck to be put on hold."

"I really wasn't sure," Nathaniel said. "The caller ID said, Unidentified Caller. I thought the Space Force guys were calling me."

"Oh, so the comm system computer hasn't updated," Adam said. "I guess we'll need to give it some time." Juni nodded acknowledgment. "But for your information, we're going to be calling our ship here the UES Conestoga. I don't know if your caller ID will say that from now on. But we probably won't be contacting you by cellphone much anymore either." He had to wait almost two minutes for the reply from Nathaniel to arrive.

"Hmm, I see," Nathaniel feigned surprise. "I'm sure you know that Ana and I have had a word or two about her delivery flight with Mila. She says you all but told her you were leaving for good. Have you had a change of heart already?"

"Oh no, not at all. That's really what I'm calling you about." Adam said. "Sorry, the delay is so long. I think I'll get out a good part of my message now so that things can go a bit faster. The thing is, this ship is just humming along fine, and nobody seems particularly homesick. You see, this is our home now! We're just as comfy as can be. So, we've decided to just work from home and take a really long vacation while we're at it. I'm panning around the room now. I don't see anybody who looks like they don't want to be here, do you? Ask anyone you want after you get this. I've sent half of the access code to the office safe. Frieda should have put Mattie's and my will in it, as well as a letter from the two of us to you. There's a letter from Kevin St. Patrick to his superiors at the Space Force declaring his intentions as well. Ana's message to you had your part of the safe code. Everything you'll need to settle our affairs is in there, so feel free to peruse it now. We'll just talk amongst ourselves while we wait for you to get this message. Let us know what you think about what's inside." Nathaniel's cell phone image, which was really just a very long-distance conference call, remains interesting to watch despite the wait time between messages. Uplift released it to the press the day after it was received. Nathaniel was there in the study, along with Frieda Jojola, Ana Kozelka, and someone Kevin recognized as a Space Force Major General stationed at Cannon Joint Air and Space Base in New Mexico. On the bridge, the crew saw the little group receive the message and go to the safe. This was inside a small closet in the study. Frieda punched in the code Adam had given her, completing the sequence Adam had already sent. Apparently, there was no time requirement between when the first and last sequences could be entered. She pulled out a large manila envelope and distributed the contents to the appropriate people. There was a brief period of waiting while the group was seen to read and digest what they were reading. The general spoke first.

"Well, Captain St. Patrick!" he said. "That's right, you've been promoted. I hear you were afraid we'd dismiss you from the service and arrest you if you ever do return to Earth. I'd have to admit that would have been my choice. It would have been so much easier for you to just communicate your intentions in the first place. I have no doubt we would have sanctioned your choice to leave if you had. This really amounts to desertion, as far as I'm concerned. But the Secretary of Defense, and the President as well, I might add, both realize that there's really nothing we can do anyway. Therefore, we all agree that you are to be congratulated for your daring by undertaking this journey. You can expect to continue to be promoted on the usual schedules so long as you make periodic reports during your journey. I guess you won't be getting paid every month, but we can put that in escrow, if you like."

"Whatever the Secretary authorizes is quite acceptable," Kevin replied, knowing the general wouldn't hear this for a few minutes. Adam chimed in as well.

"I'd like to say also that the Conestoga will be automatically sending out log entries. One every week, or sooner if an urgent situation occurs before that time. We have constructed communication satellites that we will deploy every two years to help boost our signals. Hopefully, there won't be any breaks in our transmissions." It was a pity that the signal couldn't be boosted now, as there was a delay in Nathaniel's response to this one.

"Every two years!" he finally shouted. "Where the hell are you guys going anyway? I saw that your course seems consistent with a trip to Tau Ceti, but that's ridiculous. That's almost twelve light years from here. That would mean you plan to be out for a few generations, at the least. I know you have some solar sails you were looking at and even an ion drive. Yes, I have spies around here I'm sorry to say. But for God's sake, this is nuts! You really have got a full-blown cult situation going on up there! Do you think you'll be seeing God or something? I thought you were kind of an atheist, Adam!" It must have been hard for Nathaniel to wait for the reply.

"I don't know, I guess I kind of am an atheist." Adam replied a few minutes later. "At least I believe in seeing some kind of concrete evidence for what I decide to believe in. This seems like the best way to collect that evidence, it's true. What you don't know is that we think we might have invented or Azaan has a functional antimatter drive. It's very rudimentary, of course. How could it not be? But our computers strongly indicate that we should be able to attain a velocity of just over half the speed of light. We can reach Tau Ceti theoretically within about twenty years. That factors in the ion drive and the solar sails we've just finished installing. We have everything aboard to keep us fat and happy. Of course, we still don't know if the planets found there might have adequate conditions to accommodate us. But if not, we'll still be young enough to come back. Plus, I expect we might generate a little new life of our own. So, there it is, Nathaniel. You can take it, or you can take it. We're already committed to this plan. None of us here can think of a reason not to keep going. So why don't you tell us what you win?" The crew watched and waited as the group in the study heard the message, then continued reading their parcels with extremely serious expressions. Nathaniel finally spoke.

"My God, just like you said. You and Mattie have given me full ownership of Uplift, as well as this house, all your cars, and your entire fortune... You do ask me to continue funding your charities…Frieda will be in charge of that…and you want me to have Ana keep working on projects to follow up on your research to send humans to the stars. Have you left any of that here?" Another wait ensued.

"As far as infrastructure and invoices go, yes, we have. Check with Frieda." Mattie answered. As far as the propulsion goes, Azann and the rest of us will be patiently evaluating how things work. We're studying how our velocities progress and how we all tolerate the acceleration."

"Not to mention how we tolerate each other!" Peter Hosokawa piped in.

"I really feel that you've kind of had an idea we were really going through with all of this from the very beginning," Adam said. I know we should sign off now, but can you tell me if I'm right? Have you just been faking all these objections since we started?" Another long wait.

"Well, yes and no," Nathaniel finally said. "I've known you all your life like I knew your father. Frances was the more practical of the two of you. More conventional, too, relatively speaking. He always seemed to want to do what was best for us poor earthlings. You always did, too, but you always had the historical perspective in mind when you did anything. Your father never did. That's where you surpassed your dad. I raised objections with you because I thought it would be a terrible shame if the two of you shared the same fate." The crew was seen to be becoming rather bored sitting at the bridge meeting table, making small talk as they were made to listen to this sentimental banter that barely concerned them. After this message, Adam began cutting things short.

"Well then, Nathaniel," he said. "I don't know if the press and the government are all in an uproar or not. It doesn't really matter now anyway, does it? Up here, all we have left to care about is each other. We all just hope you all don't go and blow up the planet or something while we're gone. You'll be updated on our progress regularly. Don't forget to write back. We'll try our best to show our good side to anyone we might find up here who's interested or to each other if we find nothing whatsoever. Think well of us, OK? Oh, by the way, tell everyone down there that I truly love the idea that a member of the Apache Nation is now the wealthiest man on Earth! Conestoga out!" With that, there were not to be any other communications in real time between the planet Earth and the United Earth Ship Conestoga. Communiques and ship's logs would be exchanged quite regularly, but long transmission times and even time distortions would eventually almost reduce these messages to the status of a kind of junk mail. Even so, as this chronicle documents, there would always be someone around to read that mail.

In one of the first installments of Adam's transmitted log entries, he defines that last conversation with his birth planet as his first full realization that all his plans had truly and emphatically been accomplished. He confessed that despite telling himself that completely separating himself and his companions from the rest of humanity was simply the next logical step in the advancement of Homo sapien achievement. He did admit that his only thought when he terminated that final call was "What the hell are you doing?" So maybe he wasn't quite as brave, or just foolish, as we might think he was. He also said that he spent the next few days recalling his own lifelong pattern of rationalizing why this flight needed to be undertaken and why he was the one best positioned to make it happen. "All I had to do," he wrote, "was just look around me and see that if I had been surrounded by these same people in a small, secluded village in a beautiful location on Earth, I'd feel just as safe and just as tranquil as I do here." The main thing, as he often said, was to conquer my own passions and see what would do the most good for the most people over time. He assumed the same exact feelings applied to his crewmates. He intended to continue to nurture these feelings in them as well as himself. It appears that he must have been at least partially successful in that objective. At this writing, there have been no reports of discord on the UES Conestoga. This is the reason that those of us here on Earth will always be fascinated with whatever we hear from that now legendary ship and why Uplift will continue to report anything we do hear to the rest of the world. It helps us all to remember that the planet Earth and the universe are an unlimited and infinitely complicated place. And everyone will always need to broaden their horizons. We cosmic toddlers are showing that we are now ready to start walking.

Chapter 18
Wait and Hurry Up

Following the initial departure from Earth orbit, stark reality would soon reassert itself as the main focus for everyone aboard. Basic necessities like maintaining the food supply, received a new heightened focus. Even though it had long been established that there would be no deficiencies in this area, it was only natural that the crew would be having second thoughts prior to taking their last, cautious step towards the point of no return. As an example, Sharon now moved to her long-term action plan, knowing that the initial planning phase was now complete. For instance, she realized that simply watering the plants and feeding the fish was no longer enough to keep things going in perpetuity. She got together with Juni to write some algorithms to help project how fish and crop populations might be expected to grow and at what rates. Calculus again! It didn't help that there were no winters and summers aboard ship, not to mention spring and fall. But creating seasonal lighting patterns and even small temperature fluctuations seemed a good way to help her native Earth plants adapt and flourish. Since there were no winters planned aboard, she proceeded under the assumption that growing seasons could conceivably be stacked together or even completely eliminated on board the ship. This would need to be explored. Sharon began rotating what crops she planted on a carefully planned schedule soon after departure. When one planter full of corn, rice, wheat, or barley was harvested, the bed was refreshed with compost and fertilizer and then immediately reseeded. In this way, it hopefully would soon get to the point that something needed to be harvested and stored on an almost weekly basis. This seemed encouraging. She knew that running out of food was a secret phobia that most of her fellow travelers tried to keep hidden. The fear of dying in a freak explosion or collision with an asteroid were close seconds. Sharon and Adam, therefore, made it a point to keep everyone informed about the foodstuff situation whenever they could.

The medical people were all kept busy reinforcing crew morale at every opportunity. Adam had assigned this task to Peter Hosokawa, and he, in turn, passed the job down to the rest of the medical staff. Far from being a fringe group of the crew, they began to aggressively monitor the crew's morale and treat it as the most vital job on the ship. If anyone seemed even slightly out of sorts or just a tad tired or depressed, they immediately asked if there was anything they could do that would help. All of them, even Kevin the Pharmacist, were on constant watch. They encouraged and participated in starting hobby groups and exercise sessions. Everyone was practically required to join. Within those first few weeks, Peter Hosokawa was advocating that the crew should start being assigned work schedules of five days weekly with two consecutive days off, barring any emergencies of course. Married or coupled family units would be assigned to share these "weekends" regardless of which days they fell on. He had volumes of research literature to back up his opinion. People needed regular rest intervals in order to function at top efficiency, he would often say. The crew voted unanimously to adopt this plan. The medical people were also aggressive in scheduling frequent checkups for everyone. Victoria vowed that if ever the crew made landfall again, no matter how long it took everyone would have perfect teeth when they arrived. How many people stuck back on Earth could say that? Whichever Doctor one talked to, they also never failed to advocate for a trip to the arboretum, where open space and the ability to relax in it was an endless source of tranquility, well-being, and a way to reconnect with home. The weightlessness available would soon end, of course. But there were plans to refill the space with a series of patios and platforms connected by a large number of platforms, staircases, and ladders. The recreation rooms, complete with computer-generated holographic experiences, turned out to be as popular as they were intended to be. All of these facilities were made a near mandatory part of life on the Conestoga. They always had been, but during the building period, there usually just wasn't enough time to utilize them. It would have been easy to have continued thinking of living aboard as a mere occupation, a workplace only, for the indefinite future. But the medical people made sure it didn't.

The main focus of everyone's life in these first few weeks remained to get the Conestoga ready to fully realize her potential. This was more than just establishing a status quo for life aboard, but for making this prep time as short as possible. The solar sails were installed within three, not two days. The impact on the ship's speed was relatively insignificant, and the work was performed by robots working without much need of supervision. But that extra day seemed like forever to the impatient crew. It did keep Iza busy, though, and the crew, in general, was reported to be very much pleased when the job was completed. Azaan engaged the ion propulsion system during this same period. Fuel consumption rates were found to be just as projected for this engine, and there appeared to be an appropriate supply to last just as long as would be needed, assuming nothing unexpected occurred. The ship's velocity edged up, though very slowly. If nothing changed after this, reaching their star would take several generations. But everyone felt fairly confident that much better days were not too far ahead. Three weeks after the initial rocket firing, Adam went ahead and called another meeting. Azaan dominated the conversation and pronounced that there was really no reason not to press forward with bringing the anti-matter drive online. "Anything else is just procrastination for the sake of it!" he insisted. "We could have just fired it up when we got to a safe range, and that was a couple of weeks ago!"

"We all just want to be sure it's safe!" Mila shot back. "You can't blame us for that!"

"All our analysis says it's good to go!" Azaan countered. "When have we ever questioned our own analysis? We all could have killed ourselves ten times a day already just building this ship. We'll have to push the button eventually. The longer we wait, the longer the trip. We should just face the fear and commit! That's my vote."

"I think we have all already voted to do this, haven't we?" Adam pronounced. No one could say he wasn't right. "I'd think that the only question is, are we physically ready, hardware-wise? You've all been verifying that we are, based on the data I've seen since we left Earth."

"So the only remaining step, "Manjeet piped in, "is to get the ship configured to accommodate constant thrust. That means working to a designated point in the mission elapsed time to rotate our pods and facilities to face forward. Then we can start the countdown."

Gregor spoke next." I think there are a few details we've neglected to include and some we still haven't realized we'll need. My main concern is the main passageway from the small wheel we're in right now and the rest of the ship. Right now, it's zero-G all the time. We tend to just float back and forth between the bow and the stern, right? I mean, there is a ladder we can grab onto to push and pull against to get where we're going. But when we light up the big engine, the bow will feel like up, and the stern will be down. We'll tend to fall down when we enter that causeway."

"No, we did think about that," Adam said. That's why we have a ladder."

"We have elevators in the big wheel spokes," Gregor said. "Don't forget that the shaft I'm talking about is longer than those. Climbing up and down, that beast will get old pretty quickly, long before we will. Anyway, I think I can rig up an elevator that has its own propulsion motor on the floor of the car. It can take us from this forward wheel all the way to the Arboretum. Thank heavens we had the foresight to plant most of the big trees and plants on the back wall there."

"Yeah, the trees didn't care where they were in zero-G. I think they will when we light up, though." Sharon agreed,

"Is there anything we should do with the big wheel elevators?" Adam interjected. "I know we thought they'd be OK when we designed them, but now we just float in through the doors and let the angular momentum take over, We usually have to flip over so our feet face the top. When we move out, we'll be pushed against the back wall, it will be hard to get out of the car again."

"Not at one G, I don't think," Mila said. "We'll just have to walk forward or backward against moderate resistance. We'll get used to it as we did with what we have now. Remember, we can never completely reproduce Earth conditions up here, no matter what we do. We just have to do our best, just like always. If it's a real problem, we'll figure out something later."

"Why can't we just have artificial gravity like they do on Star Trek?" Jia half giggled.

"I still like the "suspended animation" idea they always used. "Vicky said." You know, from "Lost in Space" or even in "Alien." We'd just be waking up about now, and we'd already be there!"

"Too bad we'd also be twenty years older!" Peter piped in. "There's really no way in hell we'd ever be able to suspend aging.

Plus, we'd starve to death and really need a bath! We non-Hollywood types have to live in the real world, um, universe or whatever."

"Right," Adam corrected the conversation's course. "Fortunately, we're not lost, and I'm hoping we don't have anything carnivores aboard except for ourselves. Gregor, if you can rig up your elevator soon without taking away from your other duties, I'd say go for it. We'll schedule the flip-over for the pods as soon as Azaan feels he's ready. We've always known we'd have to adjust a few things all over the ship manually when we fire up. It'll give us things to do, right?"

"Right, you are!" Mattie agreed. "Now, I assume this engine will give us a kick in the ass as well, like the last burn did?"

"Probably a bit worse initially, and the sensations may last a bit longer," Mila speculated. "It's never been done before, remember. But after that initial shock, we'll be gradually accelerating at a constant rate for years to come. We'll get used to that pretty rapidly."

"So, what I'm actually asking is, should we go back up to the Givens and strap in, or what?" Mattie wanted to know.

"Better safe than sorry," Adam decided. "I think I'd like Azaan up here with me on the bridge, though." Azaan quickly agreed. "You know, I've had it in the back of my mind that we'll need to reconfigure this bridge as well. We'll get plastered to the back wall there when we light it up. But then again, if we move things now, it'll be just as uncomfortable."

"Good point," Manjeet agreed. "But you know, I could at least fix up some seats that could support us sitting on the wall. We could use our laptops to start the burn. The gravity in here's pretty wonky as it is. Really, the whole center section of this wheel will have to be reconfigured eventually."

"OK. Do what you can before we commit," Adam said. "Moving's a bitch, isn't it."

"Aye, Captain!" Manjeet quipped. It appeared that Manjeet had decided that mimicking Scottie from Star Trek would be a good idea at this stage of preparations. His attempt at a Scottish accent was not exactly perfect, but it was oddly endearing. With no other outstanding concerns coming to mind, Adam adjourned the meeting. Mattie stayed a bit longer to develop a tentative assignment schedule for everyone, hoping to expedite the preparations leading toward this most important of countdowns. It was thought best to assign Gregor to supervise the reconfiguration of the interior spaces. As we all recall, all interior spaces of the habitation pods, including the aquariums, were built to rotate internally. This was so that the exit doors would switch from their current service corridors to the other one located 90 degrees inboard. Rotation of the wheels would no longer be necessary to hold one to the floor. The constant rearward acceleration forces would then take over. This would no doubt at least be good news for everyone's inner ears. To rotate all the habitation pods, one flick of a switch on the bridge would activate all the rotation motors to switch on simultaneously throughout the ship. This massive change in the simulated gravity would no doubt be universally disorienting. But at least there would be no need to change all these altered arrangements again for quite a few years. When the time came to slow down, the entire ship would simply wheel around 180 degrees, and the forces would wheel around with it. There would be no need to restart the wheel rotation until the ship found another celestial body to orbit. All of this was totally hypothetical at this stage. But then, so was everything else about this mission that wasn't happening at this particular instant or hadn't happened already. Right now, the crew was wondering if there would ever be a time in their lives when they weren't frantically, hysterically busy. Such was life in the big universe.

About two days into this conversion process, Azaan presented the crew with the first big challenge they would meet without any hope or prayer of any assistance possible from outside. Azaan called the bridge and reported that the x-ray camera that Sparky was to use to do a final inspection of the proton conduit of the antimatter drive was somehow not functioning. Though the 50-meter shaft had been inspected both during the manufacture of this component and again just after installation, Azaan felt one last peek at the interior of this borophene tube would be vital to maintain everyone's confidence before the damned thing was lit up for continuous use. The interior of this tube was only just over a foot wide, so a small camera was all that was needed. Iza was called in to work with Sparky to show Azaan what the bot was seeing when examining the tube. She then asked Sparky to examine the camera and the probe it was mounted on with his robotic eyes. It was soon discovered that one of the small rubber castors that helped the probe run up and down the shaft had somehow been bent out of position. Sparky lacked the dexterity to straighten the bend, but since the whole assembly had already been exposed to too much radiation to bring it back to the crew areas for repair, another solution needed to be found. Mary was called on to bring up the plans for the probe and then manufacture an entirely new one from scratch. Plans for every component of the ship were entered into the database long before and were also fully integrated with the 3D printer. The order was completed within a matter of hours. The big question remaining was how to get the part out past the radiation barrier so that Sparky could change it out with the broken one without contaminating anything in the process. The contingency plan for this was to have one of the forward bots come to an airlock, pick up the part, and then take it to just behind the crew side of the barrier. The rear area robotic arm could then pluck it away from the hands of the courier bot, Blackbeard in this case. The robot arm would then hand it off to Sparky, who could then plug in the X-ray camera and shove the whole assembly into the proton conduit shaft. This procedure was no big deal for a robotics genius like Iza, though. Iza was able to get everything back on schedule with a loss of only about 36 hours. Since the timing of the antimatter engine test was in no way essential, this turned out to be the only real bump in the process. No imperfections or cracks in the shaft were detected after a very rigorous analysis.

At the conclusion of a full seven days of rearranging, testing, cleaning, and fussing, Adam declared that he was giving the crew a two-day break with only very rudimentary necessary work to be allowed. After that, a final crew meeting would be held to serve as a basic go-no-go check list from all ship's departments. If all was, in fact, at go status, the crew would get squared away and secure, and the countdown would proceed. These two days were memorable mostly because relaxation at this point seemed impossible, especially for this perpetually wired-up collection of high energy characters. Sandra and Kevin were called on to hand out antianxiety medications even during this time off. Everyone also cursed Adam for not being very liberal about providing alcoholic beverages. Instead, he held out the promise to begin to allow the brewing of such refreshments soon after a successful firing of the big engine. Another reason that this two-day interval became memorable, at least in log entries later received at Uplift, was that this period was used behind the scenes to establish several pairings of crewmates in romantic endeavors. Stress often seems to have this kind of effect on people. Such parings were common in wartime, for example, like when men and women in war zones felt they were likely to not survive the next battle. No one actually said it, but it was also highly possible that the crew wouldn't survive the firing of this antimatter engine. This period may have been seen as everyone's last chance for love and romance. But for whatever reason, love and romance occurred in abundance. From what we have heard of future births aboard the Conestoga though, it seems that no crew additions were directly the result of this particularly stressful situation. But that would be giving too much away.

This transition period was also notably stressful for the obvious reason that rearranging everything aboard this rather large ship was quite disorienting for everyone. Cameras were generally present in all accessible spaces to make a record of where everything ended up. And sure enough, it had been a good idea. It was all but impossible not to lose track of where things had gotten to inside your pod after it had been rotated. Moving days are always harder than you think they'll be. Then, try to imagine how you would feel if you thought your new home might blow to pieces while you're trying to move. These people were either really brave or really crazy. But they got the job done on schedule.

Just like all the other milestones in this unlikely saga, the day dawned when the final hurdle in the race would have to be jumped on a track that had never been run before. But precisely how to jump it had been contemplated at length by a group of experts possessing a massive, collected body of intelligence. So it was that on June 21st, 2088, all seventeen souls aboard the Conestoga gathered on the bridge, which was currently spinning slowly around a central axis that ran from the stem to the stern of this already swift makeshift spacecraft. The layout of the bridge had been changed so that what was now considered a back wall would soon become the effective floor. What had served as a sort of observation window at the front of the bridge would soon become more of an overhead skylight. If you wanted to look straight ahead of the ship tomorrow morning, it was better to use the big flatscreen. Of course, the view was not expected to change much in the short term. As promised, Gregor had built a rather odd console seat for Adam so that he could sit anchored to the back wall, but the chair would feel like it had rotated in place 90 degrees when the wall became the floor. Before the burn, the room was basically weightless. No one truly knew which way was up when visiting the bridge. Azaan reassured everyone that when the new engine was activated, the only really significant jolt would occur initially but would feel firm and smooth thereafter. This subsequent gentle acceleration would continue for years. Despite a request from Adam that everyone have at least a small breakfast before reporting here this morning, Adam was the only one who stated that he had followed this advice. Everyone else had had enough experience with launch days to prefer empty stomachs. They knew they could eat if they were still alive after the big event. After a short discussion, it was decided to repeat the arrangements from the Earth departure burn, using the Ed Givens still docked to the outside of the big ring. Mattie, with her usual foresight, had never ordered the cargo bay seats to be removed. Simulations indicated that all the Shrikes would survive the sudden increase in acceleration without their docking latches becoming damaged or detached. If something went wrong, the Shrikes might be lost. But examinations of the latches seemed to show no problems whatsoever. So once again, the time came when everyone looked around and simply knew that the inevitable had now arrived. There would be no benefit from waiting a minute longer. There was nothing left to do but push the pedal to the floor. They would soon be traveling faster than any living being from Planet Earth ever had before. And as they all had throughout their time with the Sustained Orbital Task Force, that same mantra flowed through every mind. "Somebody's got to do it!" Adam gave the nod from there on the bridge, and they floated out. Down the main shaft, into Gregor's new elevator and back to the rear wheel. Transfer to the wheel elevator. Up and into the access corridor, then into the airlock connected to the Shrike called the Ed Givens. They filed through the Shrike airlock and back into the cargo bay, or the cockpit for Mattie, Manjeet, and Juni. Strap yourself in. Turn on the flat screens and check the Shrike's status. Call everything in on the ship's conn circuit. Listen to Adam. Oh God, the countdown is much shorter this time. No course changes happening here, just the forward thrust will be different, way different!

"Shut up, Adam!" Adam tells himself. His mind screams for him to stop right this second! But he keeps going! Why is he trying to sound so dramatic? Who knows. But he keeps on going! "Ten, nine, eight, seven, six, five, four, three, two, one, Ignition!" This first ignition would really just be a test. Just one antiproton shot out into a stream of conventional protons flowing rapidly enough so that the explosion will occur in the bell of the motor as the ship moves ahead of the massive bang. This is really how any rocket moves forward, no matter what caused the explosion in the first place. In theory, this particular bell should be up for the challenge. At least, the simulation predicts that it should be, but it will require a ten G thrust forward for a few seconds to prove it. There will be no noises associated with the test. We'll all just feel it. Here we go! Whoa! What a kick in the ass! Juni reports that the computer says the ship is coping with the test just as expected. Adam pipes in again.

"All systems are go! Computer recommends we proceed with full antiproton flow at our discretion!"

"I concur, Adam!" Azaan reports. He is monitoring things on his laptop.

"Alright, everyone!" Adam said. "I will authorize the flow of one antiproton every two seconds beginning in T-minus ten seconds. Go ahead and stay secured until I say otherwise!"

"Every two seconds?!" Azaan shouted. "I thought we had agreed on every one second!"

"Last minute decision on my part!" Adam shouted back. "My inner voice won out. Let's see how she does for twenty-four hours and then go to full power if she's happy. It won't slow us down that much."

"Ok, we'll be at .5 Gs until then, not one!" Azaan warned. Ten seconds later, another kick in the butt, but much less severe. Everyone's blood was pulsating forcefully with exhilaration. Azaan wasn't satisfied yet, however. He decided to push it. "Adam, we've been through this time and time again! We've analyzed all this in perpetuity for over two months now! The ship can tolerate full acceleration, no problem. We have more than enough fuel! Also, no problem! The only problem is the lack of will to use what we've created. We're still here, still breathing! Just keep your promise and throttle up! Trust me!" Adam apparently thought it over, but not for very long.

"Roger that! Going to full recommended antiproton flow." No countdown ensued, he just did it. Another nice kick in the ass, and this one seemed even less violent. Another two or three minutes, and there was no sensation other than the feeling that everyone felt like they were somehow feeling more substance to their bodies than they had felt since their last trip to Earth. Like most astronauts returning to a one G environment after a prolonged absence, the blood kind of flowed out of their heads and decided to flow elsewhere. Granted, in this case, it flowed backward rather than downwards, but this would all change as soon as they unbuckled and started to move around. Adam was heard to commence his stand-down sequence. Then he gave the order. "All personnel, you are free to unfasten your restraints and make your way at your leisure back to your duty stations. Report to me from there. Congratulations everyone! We've made a successful commencement of ultra-high velocity space flight!" He could be heard to be just sort of making up his phraseology as he went along. This wasn't just a burn. Burns had finite durations. This engine would be firing constantly for years on end if all went as planned. This was a commencement. And it was true that when compared to everything that had come before this moment, they were now traveling at Ultra-High Velocity. It wasn't light speed, but it was pretty damned fast. It would take Mila a few days to even try to figure out what their true velocity was. And that speed would be changing constantly as they continued to accelerate for years to come, at least until they began to decelerate years from now.

That said, the crew must have been both very proud and very humbled by the achievement. So had all those who broke records and boundaries in the past history of human flight. As soon as Adam could file a log entry and transmit it back to Earth, most humans would be able to recall where they were and what they were doing when they first heard the remarkable news. Adam Thorne had been a household word since he first took over at Uplift, despite the fact that he tried his best not to be. Most Americans at least had heard of Conestoga wagons before as well. This new Conestoga was obviously a whole different kind of wagon. But as of this writing, it still inspires just as high a level of fascination to those it left behind. That fascination has yet to wane, just as Adam had no doubt expected. The unfortunate part is that he is no longer around to see it for himself. The world is still just as fascinated by the people he took with him, people from a large assortment of nations and cultures. There is little doubt that if he had room enough aboard that ship of his, he would have taken along someone from every country on Earth. But everyone in every country is a child of this Earth. So, in a sense, he and they have succeeded in representing all of us as they move towards a star we have never seen up close with our human eyes. When and if they do, we will eventually hope to hear everything they have to tell us about what they find. Many of us will have passed away before that day. The reports might even be made by the children or even grandchildren of those few who flew away. But for all intents and purposes, it seems ordained that the journey will be completed eventually.

It will also be long debated whether or not the journey should ever even have been attempted or whether inflicting our humanity upon anything but our own world is a good idea in the first place. But the matter is now settled. The attempt is underway. As Adam always said, if this crew hadn't done it, some other crew would. And from what I know of Adam Thorne, this extremely complicated matter is in pretty good hands. As humans go, he is neither a conqueror nor a destroyer. He is just an observer, and an extremely curious one at that. So far, it appears that the team's efforts to try to narrate their journey as best they can using long-range laser technology have been successful. Uplift receives log entries from the Conestoga on an almost daily basis, even though the distances traveled by the transmissions grow longer every time they're sent. I have been compiling and occasionally publishing them since retiring from Uplift in 2104. I have tried to do the best I could in fulfilling Adam's stated goals for Uplift Corporation since he left. I believe my replacement will continue to catalog these transmissions or will at least designate someone at the organization to do it for him. I only wish that my efforts here in the realm of authoring historical literature were as competent as my abilities in the world of aerospace engineering. As I have now reached the age of 86, I am now seeking a documentarian who can compile more of those Conestoga log transmissions in order to compile the book that must be written when and if the ship ever does reach a planet in orbit around Tau Ceti and undertakes to study it. If they do anything other than turn around and come back, then their efforts there should continue to be documented for the public to peruse. That too could fill a volume or two. That doesn't even address the other efforts to explore other suns and their planets begun here at Uplift, as well as those currently being considered by other companies and nations. Adam often put his own efforts in the context of human history as a whole. He loved to compare himself to the barnstormers of the early 20th century who rarely knew what the hell they were doing at the time that they were doing it. Soon enough, antimatter propulsion itself will be seen as a sort of quaint technology that was used only in simple pursuits by extremely quaint human beings. As for me, I feel lucky to have seen what I did see happening right before my eyes and actually tried to hinder just a tad. I have tried my best to make up for any such mistakes by trying to tell this tale in the guise of a detached observer.

With Great Sincerity,

Nathaniel Floatingfeather ,

Truth or Consequences, NM

Appendix A

Crew of UES Conestoga at the time of spacecraft commissioning. Basic biographical data from Uplift Human resource files (In alphabetical order):

Tariq al Abdullah: Age at departure: 30 years. Height: 5 ft 11in, Weight: 195 pounds. Place of birth: Marrakesh Morocco. Degree: Doctor of Medicine, Residency for Neurosurgery from University of Tripoli.

Sandra Anne Annison: Age at departure: 32 years, Height: 5 ft 4in, Weight:135 pounds, Place of birth: Chicago, Illinois, USA. Degree: Doctor of Medicine from Loyola University, Chicago.

Mary Cynthia Dupree: Age at departure: 25 years, Height: 5 ft 6in, Weight:160 pounds, Place of birth: Detroit Michigan. Degree: PhD in Material Engineering from the University of New Mexico. Number Nine Member.

Jeter Hamm: Age at departure: 29years, Height 5 ft, 10in, Weight 182 pounds, Place of birth, Oslo, Norway. Degree: PhD in Nuclear Engineering from University of New Mexico. Number Nine Member after Adam Thorne had left the University.

Awamila Hayat: Age at departure: 28 years, Height 5 ft 5 in, Weight: 118 pounds, Place of birth: Chaman, Pakistan, Degree: PhD in Theoretical Astrophysics from the University of Karachi.

Peter Hosokawa: Age at departure: 32 years, Height 5 ft 8in, Weight 150 pounds, Place of birth: Kobe Japan, Degree: Doctor of Medicine from UCLA, Residency for General and Vascular Surgery from UCLA.

Sharon Layton: Age at departure: 25 years. Height 5 ft 2in, Weight 130 pounds, Place of birth, Albuquerque, New Mexico, USA, Degree:

PhD in Agriculture and Animal Husbandry from New Mexico Technical Institute.

Gregor Levinski: Age at departure: 27 years, Height: 6 ft 1in, Place of birth, St. Petersburg, Russia, Degree: PhD in Aeronautical Engineering from University of New Mexico. Number Nine Member.

Iza Mazurkiewicz: Age at departure: 28 years, Height: 5 ft 2in, Weight:140 lbs, Place of birth, Philadelphia, PA., Degree, Master of Science in Robotics from University of New Mexico. Number Nine Member.

Azaan Nazari: Age at departure: 27 years, Height: 6 ft 3in, Weight: 230 lbs, Place of birth: Mwanza, Tanzania, Degree: PhD in Rocket Propulsion from University of New Mexico, Number Nine Member.

Matilda Orona Thorne: Age at departure, 27 years, Height 5 ft 5in, Weight: 122 lbs, Place of birth, Albuquerque, NM., Degree: PhD in Aeronautical Engineering from University of New Mexico, Number Nine Member.

Junaki Sato: Age at departure: 26 years, Height 5 ft 4 in, Weight 118 lbs, Place of birth, Las Cruces, NM, Degree: PhD in Computer Engineering from University of New Mexico, Number Nine Member.

Manjeet Singh: Age at departure: 27 years. Height: 6 ft 2in, Weight:245 lbs, Place of birth: Espanola, NM, Degree: PhD in Aeronautical Engineering from University of New Mexico, Number Nine Member.

Kevin St. Patrick: Age at departure 29 years. Height: 5 ft 10in, Weight: 155 lbs, Place of birth: Washington DC, Degree: Master of Science in Pharmacology from Johns Hopkins University. Promoted to rank of Captain at commencement of UES Conestoga deep space mission.

Adam Thorne: Age at departure: 27 years. Height 5 ft 11in, Weight: 185 lbs. Place of birth: Las Cruces, NM, Degree: Master of Science in Aeronautical Engineering from University of New Mexico, Number Nine Member. Chief Executive officer and Owner of Uplift Aeronautics in Truth or Consequences New Mexico.

Victoria Suzette Zasco: Age at departure: 28 years. Height 5 ft 4in, Weight 126 lbs. Place of birth: Tuscon Az, Degree: Doctor of Dental Science, Notre Dame University, USAF Major (Retired).

Xiang Xoa Jia: Age at departure: 29 years, Height: 5 ft 3in, Weight: 118 pounds, Place of birth: Zengshou, China, Degree: Doctor of Medicine, Residency in Obstetrical and Gynecological Surgery from University of Zengshou.

(Note: This summary represents all biographical information available per OSHA regulations from Uplift Aeronautics Corporation without expressed permission from the individuals named. Such permission has not been obtained, and all individuals named are presumed to currently be living at the time of publication of this book.)

Appendix B
Timeline of Pertinent Events

September 19, 2034 - Frances Thorne born in Las Cruces New Mexico.

August 2049 - Frances enrolled as Freshman at Las Cruces High School.

May 2050 - Frances hired as janitor at Uplift Aeronautics in Truth or Consequences NM.

May 2052 - Frances graduates from high school. Receives full scholarship to attend Massachusetts Institute of Technology.

August 2052 - Frances begins Freshman year at MIT.

May 2056 - Frances earns baccalaureate in Aeronautical Engineering from MIT.

May 2057 - Frances earns master's degree at MIT in Aeronautical Engineering.

May 2059 - Frances earns PhD in Aerospace design from MIT. Patents are granted to him for design of Shrike aerospace transport vehicle. Frances sells stock and founds Uplift Aerospace Inc.

September 3rd, 2061 - Production plants completed for Shrike manufacture and headquarters and flight facilities finished at former Spaceport America in Truth or Consequences NM.

September 10th, 2061 - Frances marries film actor Verity Lovato.

December 23, 2061 - Son Adam Thorne is born in Las Cruces NM.

January 2068 - Verity divorces Frances and ceases all contact with Frances and Adam.

2067-2079- Adam attends public school at Uplift Aeronautics. Forms close relationships with schoolmates Mangeet Singh and Junaki Sato.

August 16th, 2079 - Adam, Manjeet, and Junaki begin classes at Thorne College of Aeronautical Engineering at the University of New Mexico. Adam's father declines scholarship on Adam's behalf and insists on paying all tuitions and fees for his education. Adam and others form the "Number Nine" study group. Matilda Orona joins the group.

May 2080 - Adam, Manjeet, Juni, and Matilda (Mattie) accepted for summer internship at Uplift. Adam makes first solo flight as a Shrike pilot. Mattie serves as Mission specialist on this flight.

May 2083 - Adam, Manjeet, and Junaki (Juni) complete bachelor's degrees from UNM.

September 2083 - Adam begins work at Uplift Aeronautics as a Shrike pilot and assistant to his father, who is CEO with full ownership of Uplift by this time. Mattie, Manjeet and Juni remain at UNM to work towards advanced degrees.

November 2083 - Adam granted Master of Science in Aerospace Design online, as well as a patent, for orbital fish aquarium built for his first solo Shrike Flight.

February 12th, 2084 - Frances Thorne killed at Uplift Headquarters facility by Nitrogen Tetroxide leak in storage shed. Adam is elevated to CEO position.

May 2084 - Adam assembles principle members of Uplift Sustained Orbital Research Task Force. The designated team name will change over time.

January 2085 - Uplift hired to assist in construction of an Orbital Hotel for the Infinite Horizons Inc. company.

May 30th, 2085 - Infinite Horizons Orbital Hotel is completed.

June 1st through August 30 2085 Adam gives Task Force members a three-month leave of absence to pursue personal leisure activities.

August 31, 2085- Marriage ceremony for Adam Thorne and Matilda Orona, as well as Mangeet Singh and Junaki Sato held in the front yard of the Thorne mansion at Uplift Aeronautics.

October 2085 - Task Force members instructed to attend EVA procedure training at NASA.

November 2085 - Adam informs FAA of intention to build Uplift Orbital Research Facility.

Early January 2086 - Construction begins for Uplift Orbital Research Facility.

Late January 2086 - Major Truss framework construction completed.

End of February 2086 - Installation of habitation pods completed.

Early April 2086 - Storage tanks, major wiring, heating, environmental systems, water storage, circulation, and processing equipment largely installed.

End of April 2086 - Construction of areas behind "big wheel" structure begins.

End of August 2086 - Rotation of both forward (Small) and aft (Big) habitation rings is permanently initiated. Permanent manning of station begins in earnest.

End of December 2086 - Crew holds first holiday feast aboard the station.

January 2087 - Crew begins planning theoretical interstellar journey. Target destination is determined.

January 2086 through May 2088 - Construction of engine components undertaken and completed. Fuel materials are sequestered and taken aboard for storage. Fitting out of station is ongoing, and crew duties are assigned. This is explained to Uplift management as a period of "research."

May 21st, 2088 - Conestoga Compact signed. Adam Thorne announces all development goals have been met, and the station will be commissioned as the United Earth Ship Conestoga. Plan for obtaining remaining fuel requirements will be implemented the following day.

May 22nd, 2088 - Flight of five Shrikes departs from Uplift to Karachi Pakistan. These remain overnight to take on required fuel.

May 23rd, 2088 - Single Shrike departs Karachi for Conestoga. This Shrike delivers single crew member and fuel supplies before returning to Uplift. When all is squared away, Conestoga departs Earth orbit.

May 27th - Farewell call placed to Nathaniel Floatingfeather which places him in charge of Uplift Aeronautics.

June 21st, 2088 - UES Conestoga engages main engine system to begin its voyage.

Appendix C
The Conestoga Compact

We, the crew of the newly christened UES Conestoga, do herein record our mutual consent to be governed by the following measures regarding what are conceived to be the governing laws and general rules of acceptable conduct for all persons living aboard said vessel or at any habitable destination that may be attained by said vessel. Nonadherence to said measures may be enforced and punished according to the articles listed below. This compact is intended to encourage harmonious relations between all crew members and discourage discord throughout the entire length of the voyage soon to be undertaken. This document will be applicable until such time as the UES Conestoga returns to its point of origin and the crew considers the ship's mission completed. Any policies or rules in the material included in the articles below which are contradictory to the United States Criminal Code shall be deemed to supersede that code.

Article I: The basis for all rules of conduct and legally allowed expectations for behavior shall be the 2087 edition of the United States Criminal Code. This publication has been uploaded to the database of the mainframe computer of the United Earth Ship Conestoga as of May 1st 2087.

Article II: The command structure for persons living aboard the UES Conestoga shall be as follows:

All leadership positions shall be filled by election just prior to the departure of the AES Conestoga just prior to her departure from Earth orbit. An election shall be held every 365 days thereafter for the positions of Captain, First Mate, and Section Chiefs. The Sections specified shall be Engineering, Medical, Food Production and storage, Recreation, and Astrophysics and Science.

The Captain of the vessel shall be the final authority in regard to matters pertaining to the daily operation of the ship. This includes matters of crew operations, navigation, and assignment of crew duties, as well as ensuring repair and maintenance of ship structures, systems, and equipment are performed as needed. The Captain shall also be empowered to perform marriages on the ship and will also receive performance reviews of crew members and section chiefs as required to ensure crew safety and security. All recommendations for crew disciplinary action or removal from assigned positions shall be channeled through and managed by the Captain. The Captain shall also be responsible for entering a daily summary of all crew activities in the ship's log and for maintaining a personal log on a daily basis whenever possible.

The First Mate shall be regarded as the second in command of the vessel. Duties of the First Mate shall include assuming command of the vessel at such time as the Captain is unable to discharge the duties of that position due to incapacitation, temporary leave, or deliberate temporary assignment of command duties to the First Mate. If the Captain becomes permanently unable to perform command duties, the First Mate shall assume the captaincy until an election for that position can be held by the crew. When acting as First Mate, this officer will have overall responsibility for overseeing daily work schedules for each crew member and verifying that the assigned duties of all crew members have been completed as ordered for the day.

The Engineering section shall be responsible for the maintenance, repair, and any necessary redesign of the ship throughout the duration of the mission. Subsections of this section shall be: 1- Ship's primary and secondary structures; life support infrastructure; water management; atmospheric infrastructure; heating and cooling infrastructure; waste management; computer and information services management; materials management and production/3d printing; robotics production and management; Shrike maintenance and operations; and propulsion related engineering.

The Medical section shall be responsible for maintaining general crew health and morale, maintenance of an on-board supply of medications, maintenance of a comprehensive medical record for each crew member, and for providing necessary care to all crew members in the following fields - General, Neurological and other Specialized Surgery, General Medicine, Dental and Vision care, Oncologic treatment, OBGYN services, Pharmacologic services, Radiological services, Psychologic care and treatment, Cardiac and other vascular catheterization, skin and wound care, renal and urologic care, and all other necessary medical care to the best ability of all Medical section members. A robotic surgery service will also be available and maintained. Spaces and equipment allotted to these services will be inspected and maintained on a daily basis by the medical staff.

The Food Production and Storage section shall be responsible for inspecting and maintaining all spaces and equipment related to the production of animal and vegetable-based, as well as synthetically produced foodstuffs. The section chief will be responsible for the training and scheduling of all other crewmembers to assist with these duties whenever their individual schedules allow or at any time when such assistance is necessary to ensure the procurement of an adequate food supply for the crew. These duties should include cleaning of all aquariums, feeding of aquatic life forms, breeding of aquatic species of animals and plants, and harvesting of water-based food sources. Care, watering, planting, and harvesting of all onboard plants and crops will also be managed by the section leader. This section will also be responsible for preserving and packaging all-natural and synthetically produced foodstuffs for long-term storage. Records shall be kept of all foodstuffs produced, and projections for long-term production goals shall be managed by this department. Records shall be kept of where and how all food packages are stored, as well as expiration dates for each individual package. Periodic inspection of all foodstuffs shall be undertaken as recommended by the section chief. Recommendations for infrastructure changes or maintenance should be referred in a timely manner to the Engineering section.

The Food Production Chief shall also be responsible for scheduling and managing periodic recreational activities for the crew. These may include, but are not restricted to, large group meals, special interest hobby groups, dances and parties, music-related activities and instruction, cooking lessons or group food production activities, and the use of electronics-based leisure activities. The section chief will coordinate the scheduled usage of those pods and spaces dedicated specifically for leisure activities. All of these responsibilities may also be delegated to another crew member by the section chief as long as that crew member is monitored as needed by the section chief.

The Astrophysics and Science section shall have responsibility for conducting all required and necessary observations of the environments around the spacecraft, as well as all areas of the universe observable with onboard equipment. Astronomical observations shall also be made with the on-board telescope and all other detection equipment available aboard the Conestoga. A dedicated computer system shall also be available at all times in the Astrophysics lab. Any observations made that may affect the safety of the crew or the operations of the spacecraft shall be communicated to the Captain as soon as possible. Efforts of this department shall be mainly focused on verification or nonverification of existing theories of the nature of the Universe or detection and investigation of any previously unknown phenomena or principles. Mapping of observed star systems, planetary bodies, or other astrological features shall also be the responsibility of this department. Other directives for required investigations or research may be issued as needed by the Captain.

Article III- Interrelations among crew members: Marriages between crew members may be authorized and performed by the Captain. No license or written form of authorization shall be required for this marriage to be considered legal, but the Captain must enter all mutually accepted marriages in the Ship's log as soon as possible after the ceremony is performed. Marriages require only the consent of all individuals united in the marriage. All parties must be considered by the majority of the crew to be of sound mind and demonstrate full knowledge of the responsibilities of marriage and the obligations to care for any offspring which may result from the union. Each partner must also be in good standing with other crew members in terms of safe and orderly conduct and completion of expected duties. By mutual consent of all signers of this Compact, there are no constraints on genders or number of participants in a marriage, assuming all participants freely agree with the arrangement. Habitation pods will be constrained to accommodate no more than 2 adults and three children under the age of 18 per two-pod apartment, although all family members should live as closely together as possible. Alternatively, no crewmember shall be held liable to enter into any marriage arrangement whatsoever if they do not choose to do so. Likewise, no female is liable to bear children without her consent. Neither is she obliged to marry the father of her children. Artificial conception may also be allowed, either with or without knowledge of the father's identity. All crewmembers will be held liable to assist in and assure the health and welfare of all persons born aboard the ship. All children shall be ensured of a full share in all food and resources available on board but also be liable to obey all laws and regulations in place aboard the ship for the duration of the flight. Every effort shall also be made to educate all children in ship operations, human history and culture, ethics, and applicable fields of science and mathematics. Access to onboard libraries shall always be allowed unless utilized for destructive purposes.

If a divorce between married individuals is desired, all that is required shall be mutual consent between all married parties. The Captain shall then enter this transaction in the ship's log. Separate lodging for the party requesting the separation shall then be arranged so long as it is available. The First Mate will not be obliged to change work assignments or make any arrangement which entitles the divorced parties to work apart from each other or otherwise maintain physical separation from each other in the course of their duties. Any aggressive or disruptive behavior between the parties at any time during the remainder of the voyage will be cause for punitive action against them as decided by the rest of the crew. Alternatively, the crew shall not consider divorce itself as an act worthy of punishment in its own right.

Article IIII- Religious Freedom. All crew members will respect the religious freedoms of all other crew members at all times so long as practices used as part of such worship are not detrimental to the crew or crew duties and safety. Whenever possible, crew duties should be scheduled so that observance of each member's religion may be practiced without hindrance whenever possible. Examples of religious activities may include, but are not limited to, public praying, attending religious services, or singing of hymns or traditional music. These rules should only be suspended at such times as crew safety or mission-related demands make such observance inappropriate, such as danger to the ship or other crew members, or vital repair needs for the ship.

The above rules are not to be understood to imply that any one member's beliefs are superior to or more worthy of authority than any other religion or belief so long as neither or both of these belief systems advocate harm or denigration of other persons or belief systems. No crew member shall be allowed to pressure or coerce others to share or participate in any faith or belief system against their will but may share such information only when requested to do so. Neither will any crewmember endeavor to imply that their own religious views are superior to anyone else's, or that other religions are inferior to their own. Participation in such actions shall be cause for censure by the Captain and may even be found to be cause for punitive measures by the crew if the offense is severe.

Article V- Crew Assignment Changes. If changes in crew assignments are necessary due to findings of negligence, dereliction of duty, or incompetence, it shall be the Captain's duty to appoint replacements until the next annual election. If it is felt that the Captain must be replaced for such reasons, a two-thirds majority of the crew may then call for a Court Marshall of the Captain. Such a proceeding shall use the United States Uniform Code of Military Justice as its governing document. A prosecutor and a defense counsel shall then be appointed from and by all crew members. The remainder of the crew shall act as jurors. Agreement of two thirds of the jurors shall be required to convict. All members of the crew besides the Captain shall then elect a new Captain.

It will be a requirement of the Captain to ensure that all present and future crew members for the length mission be trained so that all crew members may, in time, be qualified to occupy all other crew positions. This is not to imply that every individual receives a full education in all sciences and receives academic degrees. Rather it shall be the goal that all crew tasks may be eventually undertaken by all crew members if required. This shall not apply to such skills as surgery or scientific research, but instruction on how to use the ship's computer banks to complete necessary tasks may be used so that various tasks may be completed in the case of incapacitation or death of other skilled crewmembers. Intensive education of children who join the crew in coming years must be vigorously undertaken. All skilled crew members will be expected to teach these skills to others on an ongoing basis.

Article VI- Punishment of Violations of Crew Rules and Regulations.

Charging Crewmembers with a Crime: Establishment of guilt for any violations shall first require a report by a crew member of a possible need for formal charges to the Captain or by two crew members to the First Mate if the Captain is the subject of the charge. If the charge is severe enough to require punishment, the Captain may pronounce a humane and appropriate sentence. A trial by jury must then be held if the decision is appealed by the defendant. Charges can be laid by the Captain, or by any member of the crew."

Two unrelated parties shall be required in order to file charges if the Captain is the subject of the charge, and such charges may be filed by the First Mate, or by two thirds of the crew if there is any conflict of interest thereafter established.

Punishments: If found guilty of any charge, punishment for those convicted shall be decided by the Captain or by whoever is designated as the judge in the associated trial. Corporal punishment is to be discouraged in all but the most serious offenses which may result in the willful death or severe injury of fellow crewmates or other sentient life forms. If necessary, confinement to a habitation pod converted to serve that purpose may be used to enforce this sentence. Construction of robots to serve as guards or custodians of the prisoner may be authorized as necessary. Other sentences may include extra duties aboard the ship or restrictions from utilizing recreational areas or activities. Areas of the ship may also be made inaccessible to the convicted person. Punishments beyond those just specified must be agreed upon by all crew members with no interest in the matters related to the charge, or for matters of crew safety, or the safety of the ship.

Article VII- Interaction with New Environments - In the event that the Crew of the UES Conestoga discovers the existence of any form of extraterrestrial life, a thorough study must be undertaken using every means available to establish what impact human contact would potentially have with such life. The environment of the planet or celestial body must be demonstrated to not be subjected to any changes or destruction as a result of such contact which would negatively impact the native life forms there. Destructive elements that may be introduced by humans may include microbes into the atmosphere, microbes harmful to flora and fauna, surface temperature alterations, harmful plants, or human behavior itself. Conditions on the celestial body may also have inverse dangers to the crew members as well. Only when such negative effects are found not to exist to a dangerous degree will exchanges between the indigenous inhabitants and the crew be considered. The crew shall make every possible attempt to not reveal themselves to any sentient inhabitants of the celestial body until it is deemed appropriate by the majority of the crew.

In no instance shall the crew consider themselves as dominant over any sentient life forms discovered on this voyage. If any life forms are found in abundance that are absolutely determined to possess the intellect and characteristics of so-called "game animals" in the new environment, it may be determined that the crew may use these animals as a food source so long as the species itself is not in danger of eradication. Plant food sources may also be used as a nutrition source so long as the plants are not found to be sentient or endangered by human utilization. This also applies to any unknown form of life that may be discovered.

If interactions with the environment or indigenous life with human beings are found to be substantially harmful to the same, the crew of the UES Conestoga shall not report the existence of any findings related to the discovery to any person on the Planet Earth until such time as the Conestoga returns to Earth. This stipulation has the precise purpose of discouraging any further interaction of the indigenous life forms with the human species in the future. At no time shall crewmembers treat any discovered life forms as subservient to themselves or exert any authority over any sentient species, with the exception of those species objectively found to be food sources that are abundantly renewable or not obviously sentient. Conversely, if no adverse interactions occur between the new environment and biological elements present on the UES Conestoga (that where present when the Conestoga left the Earth), then those elements may begin to be introduced into the new environment with the goal of materially sustaining the crew for an extended period. No biologic or inorganic materials may subsequently be introduced on the planet if such action would be harmful to the new planet's ecosystem, or if it would be harmful to any inhabitants already living there.

In Conclusion, we the undersigned Crewmembers of the United Earth Ship Conestoga do hereby pledge, of our own free will, to designate the above document to serve as our preferred instrument of conduct and governance for the duration of our voyage, and our lives until such time as the Conestoga returns to the Planet Earth, and we are thereafter discharged from service aboard her.

Tariq al Abdullah, Sandra Anne Annison, Mary C. Dupree, Jeter Hamm, Awamila Hayat, Peter Hosokawa, Xiang Xoa Jia, Sharon Layton, Gregor Levinski, Iza Mazurkiewicz, Azaan Nazari, Matilda Orona, Junaki Sato, Manjeet Singh, Adam Thorne, Victoria S. Zasco, Kevin St. Patrick.

A compilation of dispatches received from the United Earth Ship Conestoga will be published in the volume "Symphony of Silence" at such time as the voyage is judged to have terminated.